OECD Principles on Water Governance

The science–policy interface is critical to the design and implementation of water policies. In theory, scientists provide policy makers with robust facts and data that can help guide decision making, and lessons from the political economy of reforms can push scientific boundaries further to trigger further research for wise solutions. While evidence-based policy is obviously desirable, in practice such a connection is not always straightforward. Another assumption behind the science–policy gap is the discrepancy between scientists and policy makers in terms of culture, process, timing, language and expected outcome.

This book tries to reconcile this discrepancy through a multi-stakeholder approach to authoring its different articles. This joint initiative between the OECD – particularly its Water Governance Initiative – and the International Water Resources Association seeks to provide a canvas for grounding water policy in science, and vice versa. The objective of this book, devoted to the *OECD Principles on Water Governance*, is to use the OECD Principles as a common thread across the articles to draw lessons from theoretical work and practical experiences in water governance reforms; but also to only feature papers authored by groups of diverse stakeholders from different institutional backgrounds.

This book was originally published as a special issue of *Water International*.

Aziza Akhmouch is the Acting Head of the OECD Division on Cities, Urban Policies and Sustainable Development. She oversees, amongst others, the OECD Water Governance Programme, which she set up in 2009 to help governments design and implement better water policies for better lives. She is the author of several publications on water governance and the founder of the OECD Water Governance Initiative, an international multi-stakeholder network gathering twice a year in a Policy Forum. She holds a PhD in Geopolitics and a MS in International Business.

Delphine Clavreul is a Counsellor in the OECD Centre for Entrepreneurship, SMEs, Regions and Cities. She was previously a Policy Analyst at the OECD Water Governance Programme. Her field of expertise covers a range of topics including multi-level governance, stakeholder engagement and water integrity. She contributed to the coordination of the OECD Water Governance Initiative, an international multi-stakeholder network sharing good practices in support of better water governance. She has contributed to several OECD water governance (country and cross-country) reports, and holds an MS in Geopolitics.

Sarah Hendry is a Senior Lecturer in Law in the Dundee Law School, and the Centre for Water Law, Policy and Science, at the University of Dundee, UK. She researches and teaches comparative legal frameworks for the regulation and governance of water resource management and water services.

Sharon B. Megdal is Professor at, and Director of, the University of Arizona Water Resources Research Center, USA. Her projects include comparison of water management policy in water-scarce regions; groundwater management and governance; managed aquifer recharge; and transboundary aquifer assessment. She is an elected Board Member for the Central Arizona Project.

James E. Nickum is an Institutional Economist affiliated to the International Water Resources Association (IWRA), France and Japan; the Centre for Water and Development at the School of Oriental and Asian Studies (SOAS), UK; the East-West Centre, Hawaii, USA; and the University of Hong Kong, China. He is Editor-in-Chief of *Water International*.

Francisco Nunes-Correia is Professor of Environment and Water Resources at IST, University of Lisbon, Portugal. He is former Minister of Environment and Regional Development of Portugal. He has been working as professor, researcher and consultant in those areas with a special interest in water policy formulation and assessment.

Andrew Ross is a Visiting Fellow and Consultant at the Fenner School of Environment and Society, Australian National University, Australia, specialising in conjunctive water management, water governance and aquifer recharge. He is a leader of the Groundwater Solutions Initiative for Policy and Practice and the IAH working group on economics of MAR.

Routledge Special Issues on Water Policy and Governance

https://www.routledge.com/series/WATER

Edited by:
Cecilia Tortajada (IJWRD) – Third World Centre for Water Management, Mexico
James Nickum (WI) – International Water Resources Association, France

Most of the world's water problems, and their solutions, are directly related to policies and governance, both specific to water and in general. Two of the world's leading journals in this area, the *International Journal of Water Resources Development* and *Water International* (the official journal of the International Water Resources Association), contribute to this special issues series, aimed at disseminating new knowledge on the policy and governance of water resources to a very broad and diverse readership all over the world. The series should be of direct interest to all policy makers, professionals and lay readers concerned with obtaining the latest perspectives on addressing the world's many water issues.

Frontiers of Land and Water Governance in Urban Regions
Edited by Thomas Hartmann and Tejo Spit

The Water-Energy-Food Nexus in the Middle East and North Africa
Edited by Martin Keulertz and Eckart Woertz

Sustainability in the Water-Energy-Food Nexus
Edited by Anik Bhaduri, Claudia Ringler, Ines Dombrowsky, Rabi Mohtar and Waltina Scheumann

Transboundary Water Cooperation
Principles, Practice and Prospects for China and its Neighbours
Edited by Patricia Wouters, Huiping Chen and James E. Nickum

Water Reuse Policies for Potable Use
Edited by Cecilia Tortajada and Choon Nam Ong

Hydrosocial Territories and Water Equity
Theory, Governance and Sites of Struggles
Edited by Rutgerd Boelens, Ben Crow, Jaime Hoogesteger, Flora E Lu, Erik Swynedouw and Jeroen Vos

Energy for Water
Regional Case Studies
Edited by Christopher Napoli

Legal Mechanisms for Water Resources in the Third Millennium
Select papers from the IWRA XIV and XV World Water Congresses
Edited by Marcella Nanni, Stefano Burchi, Ariella D'Andrea and Gabriel Eckstein

Integrated Water Management in Canada
The Experience of Watershed Agencies
Edited by Dan Shrubsole, Dan Walters, Barbara Veale and Bruce Mitchell

Groundwater and Climate Change
Multi-Level Law and Policy Perspectives
Edited by Philippe Cullet and Raya Marina Stephan

OECD Principles on Water Governance
From Policy Standards to Practice
Edited by Aziza Akhmouch, Delphine Clavreul, Sarah Hendry, Sharon B. Megdal, James E. Nickum, Francisco Nunes-Correia and Andrew Ross

OECD Principles on Water Governance

From Policy Standards to Practice

Edited by

Aziza Akhmouch, Delphine Clavreul, Sarah Hendry, Sharon B. Megdal, James E. Nickum, Francisco Nunes-Correia and Andrew Ross

IWRA

OECD

Routledge
Taylor & Francis Group
LONDON AND NEW YORK

First published 2019
by Routledge
2 Park Square, Milton Park, Abingdon, Oxon, OX14 4RN, UK

and by Routledge
711 Third Avenue, New York, NY 10017, USA

Routledge is an imprint of the Taylor & Francis Group, an informa business

British Library Cataloguing in Publication Data
A catalogue record for this book is available from the British Library

ISBN 13: 978-1-138-32976-8

Typeset in Minion Pro
by RefineCatch Limited, Bungay, Suffolk

Publisher's Note
The publisher accepts responsibility for any inconsistencies that may have arisen during the conversion of this book from journal articles to book chapters, namely the possible inclusion of journal terminology.

Disclaimer
Every effort has been made to contact copyright holders for their permission to reprint material in this book. The publishers would be grateful to hear from any copyright holder who is not here acknowledged and will undertake to rectify any errors or omissions in future editions of this book.

Contents

Citation Information

The chapters in this book were originally published in *Water International*, volume 43, issue 1 (January 2018). When citing this material, please use the original page numbering for each article, as follows:

Foreword

Editors' foreword
Aziza Akhmouch, Delphine Clavreul, Sarah Hendry, Sharon B. Megdal, James E. Nickum, Francisco Nunes-Correia and Andrew Ross
Water International, volume 43, issue 1 (January 2018), pp. 1–4

Introduction

Introducing the OECD Principles on Water Governance
Aziza Akhmouch, Delphine Clavreul and Peter Glas
Water International, volume 43, issue 1 (January 2018), pp. 5–12

Chapter 1

Addressing the policy-implementation gaps in water services: the key role of meso-institutions
Claude Ménard, Alejandro Jimenez and Hakan Tropp
Water International, volume 43, issue 1 (January 2018), pp. 13–33

Chapter 2

Stakeholder engagement in water governance as social learning: lessons from practice
Uta Wehn, Kevin Collins, Kim Anema, Laura Basco-Carrera and Alix Lerebours
Water International, volume 43, issue 1 (January 2018), pp. 34–59

Chapter 3

OECD Principles on Water Governance in practice: an assessment of existing frameworks in Europe, Asia-Pacific, Africa and South America
Susana Neto, Jeff Camkin, Andrew Fenemor, Poh-Ling Tan, Jaime Melo Baptista, Marcia Ribeiro, Roland Schulze, Sabine Stuart-Hill, Chris Spray and Rahmah Elfithri
Water International, volume 43, issue 1 (January 2018), pp. 60–89

Chapter 4

Functions of OECD Water Governance Principles in assessing water governance practices: assessing the Dutch Flood Protection Programme
Chris Seijger, Stijn Brouwer, Arwin van Buuren, Herman Kasper Gilissen, Marleen van Rijswick and Michelle Hendriks
Water International, volume 43, issue 1 (January 2018), pp. 90–108

Chapter 5

The evolution of water governance in France from the 1960s: disputes as major drivers for radical changes within a consensual framework
Marine Colon, Sophie Richard and Pierre-Alain Roche
Water International, volume 43, issue 1 (January 2018), pp. 109–132

Notes on Contributors

Aziza Akhmouch is Acting Head of the Division on Cities, Urban Policies and Sustainable Development at the Organisation for Economic Co-operation and Development (OECD), Paris, France.

Kim Anema is a Lecturer in Flood Resilience in the Integrated Water Systems and Governance Department, IHE Delft Institute for Water Education, The Netherlands, and Cofounder of De Nieuwe Vrijwilliger, Nijmegen, The Netherlands.

Laura Basco-Carrera is the International Water Resources Management Group Coordinator at the Water Youth Network, Delft, The Netherlands.

Stijn Brouwer is a Senior Researcher at the KWR Water Cycle Research Institute, Nieuwegein, The Netherlands.

Arwin van Buuren is Endowed Professor of Public Administration at Erasmus University, Rotterdam, The Netherlands.

Jeff Camkin is Professor of Water Resources Management at the Centre of Excellence in Natural Resource Management, University of Western Australia, and a Visiting Professor at the Laboratório Nacional de Engenharia Civil, Lisbon, Portugal.

Delphine Clavreul is a Counsellor in the Centre for Entrepreneurship, SMEs, Regions and Cities at the Organisation for Economic Co-operation and Development (OECD), Paris, France.

Kevin Collins is a Senior Lecturer in Environment and Systems in the School of Engineering and Innovation at The Open University, Milton Keynes, UK.

Marine Colon is a Teacher and Researcher at AgroParisTech, Montpellier, France.

Rahmah Elfithri is a Senior Lecturer and Research Fellow at the Institute for Environment and Development (LESTARI), Universiti Kebangsaan Malaysia (UKM). She is also coordinator of the UNESCO-IHP HELP for Langat River Basin in Malaysia.

Andrew Fenemor is a Senior Scientist in Integrated Catchment Management at Landcare Research, Nelson, New Zealand. He is coordinator of the Motueka catchment ICM and HELP programme, and is a water commissioner for decision-making under New Zealand's Resource Management Act.

Herman Kasper Gilissen is Assistant Professor in the Department of Law, Economics and Governance and the Utrecht Centre for Water, Oceans and Sustainability Law at Utrecht University, The Netherlands.

Peter Glas is Chairman of the Water Board De Dommel, The Netherlands, and Chair of the OECD Water Governance Initiative.

Michelle Hendriks is affiliated with the Dutch Flood Protection Programme, Utrecht, The Netherlands.

Sarah Hendry is a Senior Lecturer in Law in the Dundee Law School, and the Centre for Water Law, Policy and Science at the University of Dundee, UK.

Alejandro Jimenez is a Programme Manager at the Water Governance Facility at the Stockholm International Water Institute, Sweden.

Alix Lerebours is a Board Member at the Water Youth Network, Delft, The Netherlands.

Sharon B. Megdal is Professor and Director at the University of Arizona Water Resources Research Center, USA.

Jaime Melo Baptista is a Principal Researcher in the Hydraulics and Environment Department at the Laboratório Nacional de Engenharia Civil, Lisbon, Portugal, and coordinator of the Lisbon International Centre for Water (LIS-Water).

Claude Ménard is Emeritus Professor at the University of Paris (Panthéon-Sorbonne), Paris, France.

Susana Neto is a Researcher with the University of Lisbon, Portugal, and an Adjunct Professor at the School of Civil, Environmental and Mining Engineering, University of Western Australia, Perth, Australia. She is President of the Portuguese Water Association.

James E. Nickum is an Institutional Economist affiliated to the International Water Resources Association, Paris, France.

Francisco Nunes-Correia is Professor of Environment and Water Resources at the Instituto Superior Técnico, University of Lisbon, Portugal.

Marcia Ribeiro is Associate Professor at the Department of Civil Engineering, Centre for Technology and Natural Resources, Federal University of Campina Grande, Campina Grande, Brazil.

Sophie Richard is the Head of the Water Management Section at AgroParisTech, Montpellier, France.

Marleen van Rijswick is Professor of European and Dutch Water Law in the Department of Law, Economics and Governance, and Director of the Utrecht Centre for Water, Oceans and Sustainability Law at Utrecht University, The Netherlands.

Pierre-Alain Roche is the President of the Transport Department at the French Ministry of Ecological and Inclusive Transition, Tour Sequoia, La Défense, France.

Andrew Ross is a Visiting Fellow and Consultant at the Fenner School of Environment and Society, Australian National University, Australia, specialising in conjunctive water management, water governance and managed aquifer recharge. He is a leader of the Groundwork Solutions Initiative for Policy and Practice and the IAH working group on economics of MAR.

Roland Schulze is Professor Emeritus at the Centre for Water Resources Research in the School of Agricultural, Earth and Environmental Sciences at the University of KwaZulu-Natal, Durban, South Africa.

Chris Seijger is a Postdoctoral Researcher in Environmental Planning and Water Governance at IHE Delft (Department of Integrated Water Systems and Governance) and Freiburg University (Chair of Forest and Environmental Policy).

Chris Spray is Professor of Water Science and Policy in the School of Social Sciences, UNESCO Centre for Water Law, Policy and Science at the University of Dundee, UK.

Sabine Stuart-Hill is a Lecturer at the School for Agricultural, Earth and Environmental Sciences at the University of KwaZulu-Natal, and a Senior Researcher at the Centre for Water Resources Research, Durban, South Africa. She closely collaborates various governmental initiatives on local up to national level.

Poh-Ling Tan is Professor of Water Law and Governance, a member of the Australian Rivers Institute and Law Futures Centre at Griffith University, Brisbane, Australia, and also the OECD Water Governance Initiative. She advises the Murray-Darling Authority and the Queensland Government on aspects of water policy.

Hakan Tropp is the Director of Water Governance at Stockholm International Water Institute, Sweden.

Uta Wehn is Associate Professor of Water Innovation Studies in the Integrated Water Systems and Governance Department at IHE Delft Institute for Water Education, The Netherlands.

Editors' foreword

Aziza Akhmouch, Delphine Clavreul, Sarah Hendry, Sharon B. Megdal,
James E. Nickum, Francisco Nunes-Correia and Andrew Ross

The science–policy interface is critical to the design and implementation of 'better policies for better lives', a motto at the core of the mission of the Organisation for Economic Co-operation and Development (OECD). In theory, scientists provide policy makers with robust facts and data that can help guide decision making, and lessons from the political economy of reforms can push scientific boundaries further to trigger further research for wise solutions

While evidence-based policy is obviously desirable, in practice such a connection is not always straightforward. Science and other forms of knowledge are not always used effectively in policy making; and policy makers do not always effectively inform scientists about their needs for scientific evidence to support their policy choices. In broad terms, the goals of scientists can be either to improve policies that affect science (policy for science) or to improve policies that can benefit from scientific understanding (science for policy). While science can generate objective and credible information, policy choices also rely on subjective value judgements to define the desirable outcomes and to manage trade-offs between competing interests. One explanation, as economics and game theory suggest, is that agents are rational in the sense that they have clear preferences, model uncertainty, and tend to perform the action with the optimal expected outcome for themselves from among all feasible actions. Under that framework, policy makers may therefore choose to go in one direction and take 'questionable' decisions, when objective facts and data suggest otherwise. But in practice, behavioural economics also recognizes cognitive limitations, heuristics, systematic biases, uncertainty and bounded rationality. This is an important dimension, as agents often have multiple, sometimes conflicting goals, and do not always behave according to their stated goals, for example because social acceptability is more important to them than material gain. Also 'facts' depend on perceptions, for example 'wasting water' means something very different for irrigators than for environmentalists. In sum, science, objective facts and data do have a role to guide better decision making in water management, but this does not always happen, for a number of reasons related to individual behaviours and the broader political economy of reform.

The opinions expressed and arguments employed herein are solely those of the author(s) and do not necessarily reflect the official views of the OECD or of its member countries.

Another assumption behind the science–policy gap is the discrepancy between scientists and policy makers in terms of culture, process, timing, language and expected outcome. This special issue has tried to reconcile this discrepancy through a multi-stakeholder approach to authoring its different articles. This joint initiative between the OECD – particularly its Water Governance Initiative[1] – and the International Water Resources Association[2] seeks to provide a canvas for grounding water policy in science, and vice versa.

This ambition stemmed from both organizations' core aspiration and shared understanding that 'governance' is not only about 'governments'. The mission of the OECD is to promote policies that will improve the economic and social well-being of people around the world. The OECD advises governments through economic analyses, data and benchmarks, international best practice and policy standards. In particular, it has supported water policy reform in countries as far-flung and varied as Mexico, the Netherlands, Jordan, Tunisia, Brazil and Korea, in addition to providing other benchmarks and thematic work. Its Water Governance Initiative has become a leading multi-stakeholder forum, gathering twice a year to share common solutions to common problems.

The International Water Resources Association, as an educational and research organization, has been promoting effective science–policy interface through information exchange across disciplines and communication between researchers and policy makers for more than 45 years. Most recently, the association organized its XVI World Water Congress, Bridging Science and Policy, which concluded with the Cancun Declaration, stating that 'Multidisciplinary knowledge, evidence based policies, involvement and participation of everybody … are needed to realise the 2030 Sustainable Development Agenda.'

Since their adoption in 2015, the OECD Principles on Water Governance have provided a framework for governments and stakeholders to promote good governance in support of better water outcomes. The principles recognize the inherent complexity of the water cycle, its vital contribution to health, poverty alleviation, agriculture and energy, and the multiplicity of actors and interests in water policy. They emphasize that water 'crises' are not objective states of nature; nor are they primarily due to the lack of knowledge or guidance on 'what to do' to fix the challenges of too much, too little or too polluted waters. They are rather associated with the poor articulation of who does what, how, and at which scale; in a nutshell, they often have to do with poor governance.

The objective of this special issue devoted to the OECD Principles on Water Governance is twofold. First, the principles are used as a common thread across the articles to draw lessons from theoretical work and practical experiences in water governance reforms. Second, the call for papers itself clearly targeted themes to be authored by groups of diverse stakeholders involved in the Water Governance Initiative, including academics, regulators, utilities, NGOs, international organizations, user representatives and policy makers. The originality of this approach lies in the value it adds to the *process* underlying the preparation of each paper, as much as to the *outcome*, which in this case is a compendium of peer-reviewed articles. Indeed, the very design of the special issue provided a platform for dialogue among groups of authors from different institutional backgrounds that confronted their opinions and experience on water governance, to work together to deliver common articles related to the principles.

This special issue begins with an introduction that recalls the rationale and process of developing the OECD Principles on Water Governance. Akhmouch et al. provide a detailed overview of the 12 principles and refer to current work being undertaken by the Water Governance Initiative to support their implementation, including the development of indicators and the stories. The five subsequent papers blend theoretical reflections on recent trends and trajectories in water governance, with practical inputs, to provide the reader with lessons learned from putting water governance in practice in different regions.

Menard et al. address the implementation 'gap' that has been endemic in water reforms. This gap is often the result of misalignment among the institutional arrangements, the incentives, and the resources mobilized in water policies. This paper highlights the role that 'meso-institutions' play in bridging the policy-making and policy-implementation levels. But meso-institutions are weakened by a number of flaws related to the policy-formulation process, policy operationalization, the behaviour of stakeholders, and the broader governance environment in which the previous factors are deeply embedded. The authors then explore how the OECD Principles on Water Governance are primarily and rightly targeting these flaws and can contribute to strengthening meso-institutions so they form a central link in the water policy implementation chain, i.e. by both facilitating the implementation of policies and channelling information and requests about the needs, resources, and social demand of users.

Wehn et al. use as a starting point the common assertion that stakeholder engagement is an integral part of sound governance processes, to explore the range of existing practices for engaging with stakeholders. They build on their respective experiences of coordinating consultation exercises on flood risk management, Integrated Water Resources Management and water security, to improve the understanding of OECD Principle no.10, 'Promoting stakeholder engagement for informed and outcome-oriented contributions to water policy design and implementation'. They conclude by suggesting a reframing of stakeholder engagement as a process of social learning that opens up more possibilities than just participation as it carries an explicit purpose that underpins design and process considerations. This requires consideration of the ethics, process, participants, roles and expected outcomes of stakeholder engagement, as reflected in the OECD principles.

Neto et al. assess the extent to which selected policy and legal frameworks on water from around the world align with the OECD principles. In particular, the authors analyze national water policies from Australia, Brazil, New Zealand and South Africa, the European Union's Water Framework Directive and the Lisbon Charter on water and sanitation to assess whether these frameworks are conducive to the implementation of the 12 OECD principles. To this end they employ the adapted criteria from the Likert scale: alignment, implementation, on-the-ground results, and policy impacts. They conclude by identifying which standards captured in the OECD principles are already well featured in the selected water policy frameworks, and which ones deserve greater consideration to improve water governance.

Seijger et al. explore the practical value of the OECD principles in assessing water governance practices by presenting a method for applying the principles as a tool for reflexive learning. The authors review the actual learning assessment of the Dutch Flood Protection Programme, which included independent and participatory methods to collect data. In doing so, they highlight various functions of the OECD principles, from enhancing understanding to reforming water policy agendas. The authors conclude that the OECD principles provide a meaningful action-oriented assessment matrix that includes contextualization, multiple methods, inclusiveness and periodic assessments, and which can be used to identify challenges in water governance systems well as well solutions to overcome them.

Finally, Colon et al. take the reader behind the scenes of more than 50 years of water governance reforms in France through the lens of the OECD principles. They analyze the pace of incremental evolution and consensus building as well as the most recent policy changes related to water resources and services management, and show that tensions among key water players have played an important role, driving policy changes and shaping the current French water policy framework, which aligns with the OECD principles.

The context of freshwater management has radically changed in the last 25 years. We must recognize that the governance 'climate' has significantly changed too, not least because better and more accessible information can shed greater light on poor practices. Today, the devastating effects of water crises on food security, poverty alleviation, economic development and social stability are forcing decision makers to make tough choices about how to manage water for inclusive economic growth and environmental stability.

Better engaging researchers and scientists can help ensure that these choices are the right ones. By documenting theoretical reflections and practical experiences of water governance in practice, we hope this special issue contributes to bridging the gap between the academic literature and practitioners; and that it inspires stakeholders to form similar partnerships across communities of practices. The knowledge thus gained will, we trust, contribute to building a two-way street between water policy making and science.

Notes

1. The OECD Water Governance Initiative is an international multi-stakeholder network of members from the public, private and non-for-profit sectors gathering twice a year to share good practices in support of better governance in the water sector.
2. The IWRA is a non-profit, non-governmental, educational organization that provides a global, knowledge-based forum for disciplines and geographies by connecting professionals, students, individuals, corporations and institutions who are concerned with the sustainable use of the world's water resources.

INTRODUCTION

Introducing the OECD Principles on Water Governance

The need for resilient governance to face current and future water risks

The intensifying competition for water resources is well documented. Both demand-side and supply-side pressures are on the rise, driven by economic development, population growth, deteriorating water quality and climate change. Projections to 2050 point out that the world population will reach 9.7 billion people and that water demand will rise by 55% (OECD, 2012a). What is more, 4 billion people will live in water-stressed river basins, including in 'water-rich' countries, as evidenced by the recent scarcity crisis between Rio de Janeiro and São Paulo in Brazil. Further conflicts will be inevitable without a serious transition from *crisis* management to *risk* management.

Coping with current and future challenges requires robust public policies, targeting measurable objectives in predetermined time schedules at the appropriate scale, relying on a clear assignment of duties across responsible authorities and subject to regular monitoring and evaluation. It raises the questions, not only of 'what to do' but also of 'who does what', 'why', 'at which level of government' and 'how'. Given that many water crises are often foremost crises of governance (OECD, 2011), the OECD argues that policy responses will be viable only if they are coherent, if stakeholders are properly engaged, if well-designed regulatory frameworks are in place, if there is adequate and accessible information, and if there is sufficient capacity, integrity and transparency.

Water governance can greatly contribute to the design and implementation of such policies, in a shared responsibility across levels of government, civil society, business and the broader range of stakeholders who have an important role to play alongside policy makers to reap the economic, social and environmental benefits of good water governance. It encompasses the range of political, institutional and administrative rules, practices and processes (formal and informal) through which decisions are taken and implemented, stakeholders can articulate their interests and have their concerns considered, and decision makers are held accountable for water management (OECD, 2015a).

Evidence shows that there is not a one-size-fits-all solution to water challenges worldwide, but rather a large diversity of situations within and across countries. Governance responses should therefore be adapted to territorial specificities, recognizing that governance is highly context-dependent and it is important to fit water policies to places. But the governance landscape of freshwater management has changed in the last 25 years. Information flows more easily and potentially sheds greater light on deficiencies, failures and poor practices. Decentralization resulted in opportunities to customize policies to local realities, but also raised capacity and coordination challenges in the delivery of public services.

In 2009, the OECD set up its Water Governance Programme to identify and help governments, at all levels, bridge critical governance gaps in the design and implementation of their water policies, through economic analysis, policy dialogues, standards and international best practices. This work has relied on the strong affirmation that water management should not be confined to the limits of a *sectoral* or *environmental* issue but be approached primarily as an *economic* issue that is decisive for sustainable and inclusive growth, territorial development and well-being at large.

The OECD Principles on Water Governance

Water management has intrinsic characteristics that make it highly sensitive to and dependent on multilevel governance. First, water connects across sectors, places and people, as well as geographic and temporal scales. In most cases, hydrological boundaries and administrative perimeters do not coincide. Second, freshwater management (surface and groundwater) is both a global and local concern, and involves a plethora of public, private and non-profit stakeholders in the decision-making, policy and project cycles. Third, water is a highly capital-intensive and monopolistic sector, with important market imperfections where coordination is essential. Also, water policy is inherently complex and strongly linked to domains that are critical for development, including health, environment, agriculture, energy, spatial planning, regional development and poverty alleviation. Lastly, to varying degrees, countries have allocated increasingly complex and resource-intensive responsibilities to sub-national governments, resulting in interdependencies across levels of government that require coordination to mitigate fragmentation.

Since 2009, the OECD Water Governance Programme has provided evidence on the main governance gaps hindering water policy design and implementation in 17 OECD countries (OECD, 2011) and 13 Latin American countries (OECD, 2012b), while supporting water reforms as part of national policy dialogues in Mexico (OECD, 2013), the Netherlands (OECD, 2014a), Jordan (OECD, 2014b), Tunisia (OECD, 2014c), Brazil (OECD, 2015b; OECD, 2017a) and Korea (OECD, 2017b). Thematic knowledge and policy guidance were also developed on stakeholder engagement (OECD, 2015c), the governance of water regulators (OECD, 2015d), and water governance in cities (OECD, 2016). This work reached its apex with the OECD Principles on Water Governance (Table 1), which were developed through a bottom-up and multi-stakeholder process within the OECD Water Governance Initiative.[1] The principles were adopted in June 2015 by the 35 OECD member countries to support effective, efficient and inclusive water policies and thus improve the 'water governance cycle', from policy design to implementation (Figure 1). In December 2016, the OECD Principles on Water Governance were included verbatim in a broader OECD Council Recommendation on Water, which also covers issues of water quantity, water quality, water risks and disasters, and sustainable finance, investment and pricing.

The principles are articulated around three mutually reinforcing and complementary dimensions of water governance (Figure 2).

- *Effectiveness* relates to the contribution of governance to define clear sustainable water policy goals and targets at all levels of government, to implement those policy goals, and to meet expected targets. It calls for clearly allocating roles and

Table 1. The OECD Principles on Water Governance.

Enhancing the **effectiveness** of water governance	
Principle 1	a) **Clearly allocate and distinguish roles and responsibilities for water policymaking, policy implementation, operational management and regulation, and foster co-ordination across these responsible authorities**
	b) Specify the allocation of roles and responsibilities, across all levels of government and water-related institutions in regard to water
	c) Help identify and address gaps, overlaps and conflicts of interest through effective co-ordination at and across all levels of government.
Principle 2	d) **Manage water at the appropriate scale(s) within integrated basin governance systems to reflect local conditions, and foster co-ordination between the different scales.**
	e) Respond to long-term environmental, economic and social objectives with a view to making the best use of water resources, through risk prevention and integrated water resources management;
	f) Encourage a sound hydrological cycle management from capture and distribution of freshwater to the release of wastewater and return flows;
	g) Promote adaptive and mitigation strategies, action programs and measures based on clear and coherent mandates, through effective basin management plans that are consistent with national policies and local conditions;
	h) Promote multi-level co-operation among users, stakeholders and levels of government for the management of water resources; and,
	i) Enhance riparian co-operation on the use of transboundary freshwater resources.
Principle 3	j) **Encourage policy coherence through effective cross-sectoral co-ordination, especially between policies for water and the environment, health, energy, agriculture, industry, spatial planning and land use**
	k) Encouraging co-ordination mechanisms to facilitate coherent policies across ministries, public agencies and levels of government, including cross-sectoral plans;
	1) Fostering co-ordinated management of use, protection and clean-up of water resources, taking into account policies that affect water availability, quality and demand (e.g. agriculture, forestry, mining, energy, fisheries, transportation, recreation, and navigation) as well as risk prevention;
	m) Identifying, assessing and addressing the barriers to policy coherence from practices, policies and regulations within and beyond the water sector, using monitoring, reporting and reviews; and
	n) Providing incentives and regulations to mitigate conflicts among sectoral strategies, bringing these strategies into line with water management needs and finding solutions that fit with local governance and norms.
Principle 4	o) **Adapt the level of capacity of responsible authorities to the complexity of water challenges to be met, and to the set of competencies required to carry out their duties**
	p) Identifying and addressing capacity gaps to implement integrated water resources management, notably for planning, rule-making, project management, finance, budgeting, data collection and monitoring, risk management and evaluation;
	q) Matching the level of technical, financial and institutional capacity in water governance systems to the nature of problems and needs;
	r) Encouraging adaptive and evolving assignment of competences upon demonstration of capacity, where appropriate;
	s) Promoting hiring of public officials and water professionals that uses merit-based, transparent processes and are independent from political cycles; and
	t) Promoting education and training of water professionals to strengthen the capacity of water institutions as well as stakeholders at large and to foster co-operation and knowledge-sharing
Enhancing the **efficiency** of water governance	
Principle 5	a) **Produce, update, and share timely, consistent, comparable and policy-relevant water and water-related data and information, and use it to guide, assess and improve water policy**
	b) Defining requirements for cost-effective and sustainable production and methods for sharing high quality water and water-related data and information, e.g. on the status of water resources, water financing, environmental needs, socio-economic features and institutional mapping
	c) Fostering effective co-ordination and experience sharing among organisations and agencies producing water-related data between data producers and users, and across levels of government;
	d) Promoting engagement with stakeholders in the design and implementation of water information systems, and providing guidance on how such information should be shared to foster transparency, trust and comparability (e.g. data banks, reports, maps, diagrams, observatories);
	e) Encouraging the design of harmonised and consistent information systems at the basin scale, including in the case of transboundary water, to foster mutual confidence, reciprocity and comparability within the framework of agreements between riparian countries; and
	f) Reviewing data collection, use, sharing and dissemination to identify overlaps and synergies and track unnecessary data overload.

(*Continued*)

Table 1. (Continued).

Principle 6	g) **Ensure that governance arrangements help mobilise water finance and allocate financial resources in an efficient, transparent and timely manner** h) Promoting governance arrangements that help water institutions across levels of government raise the necessary revenues to meet their mandates, building through for example principles such as the polluter-pays and user-pays principles, as well as payment for environmental services; i) Carrying out sector reviews and strategic financial planning to assess short, medium and long term investment and operational needs and take measures to help ensure availability and sustainability of such finance; j) Adopting sound and transparent practices for budgeting and accounting that provide a clear picture of water activities and any associated contingent liabilities including infrastructure investment, and aligning multi-annual strategic plans to annual budgets and medium-term priorities of governments; k) Adopting mechanisms that foster the efficient and transparent allocation of water-related public funds (e.g. through social contracts, scorecards, and audits); and 1) Minimising unnecessary administrative burdens related to public expenditure while preserving fiduciary and fiscal safeguards.
Principle 7	**Ensure that sound water management regulatory frameworks are effectively implemented and enforced in pursuit of the public interest** a) Ensuring a comprehensive, coherent and predictable legal and institutional framework that set rules, standards and guidelines for achieving water policy outcomes, and encourage integrated long-term planning; b) Ensuring that key regulatory functions are discharged across public agencies, dedicated institutions and levels of government and that regulatory authorities are endowed with necessary resources; c) Ensuring that rules, institutions and processes are well-co-ordinated, transparent, non-discriminatory, participative and easy to understand and enforce; d) Encouraging the use of regulatory tools (evaluation and consultation mechanisms) to foster the quality of regulatory processes and make the results accessible to the public, where appropriate; e) Setting clear, transparent and proportionate enforcement rules, procedures, incentives and tools (including rewards and penalties) to promote compliance and achieve regulatory objectives in a cost-effective way; and f) Ensuring that effective remedies can be claimed through non-discriminatory access to justice, considering the range of options as appropriate.
Principle 8	g) **Promote the adoption and implementation of innovative water governance practices across responsible authorities, levels of government and relevant stakeholders** h) Encouraging experimentation and pilot-testing on water governance, drawing lessons from success and failures, and scaling up replicable practices; i) Promoting social learning to facilitate dialogue and consensus-building, for example through networking platforms, social media, Information and Communication Technologies (ICTs) and user-friendly interface (e.g. digital maps, big data, smart data and open data) and other means; j) Promoting innovative ways to co-operate, to pool resources and capacity, to build synergies across sectors and search for efficiency gains, notably through metropolitan governance, inter-municipal collaboration, urban-rural partnerships, and performance-based contracts; and k) Promoting a strong science-policy interface to contribute to better water governance and bridge the divide between scientific findings and water governance practices.
Enhancing **trust and engagement** in water governance	
Principle 9	a) **Mainstream integrity and transparency practices across water policies, water institutions and water governance frameworks for greater accountability and trust in decision-making** b) Promoting legal and institutional frameworks that hold decision-makers and stakeholders accountable, such as the right to information and independent authorities to investigate water related issues and law enforcement; c) Encouraging norms, codes of conduct or charters on integrity and transparency in national or local d) Diagnosing and mapping on a regular basis existing or potential drivers of corruption and risks in all water-related institutions at different levels, including for public procurement; and e) Adopting multi-stakeholder approaches, dedicated tools and action plans to identify and address water integrity and transparency gaps (e.g. integrity scans/pacts, risk analysis, social witnesses)

(*Continued*)

Table 1. (Continued).

Principle 10	f) **Promote stakeholder engagement for informed and outcome-oriented contributions to water policy design and implementation**
	g) Mapping public, private and non-profit actors who have a stake in the outcome or who are likely to be affected by water-related decisions, as well as their responsibilities, core motivations and interactions;
	h) Paying special attention to under-represented categories (youth, the poor, women, indigenous people, domestic users) newcomers (property developers, institutional investors) and other water-related stakeholders and institutions;
	i) Defining the line of decision-making and the expected use of stakeholders' inputs, and mitigating power imbalances and risks of consultation capture from over-represented or overly vocal categories, as well as between expert and non-expert voices;
	j) Encouraging capacity development of relevant stakeholders as well as accurate, timely and reliable information, as appropriate;
	k) Assessing the process and outcomes of stakeholder engagement to learn, adjust and improve accordingly, including the evaluation of costs and benefits of engagement processes;
	1) contexts and monitoring their implementation; Establishing clear accountability and control mechanisms for transparent water policy making and implementation;
	n) Promoting legal and institutional frameworks, organisational structures and responsible authorities that are conducive to stakeholder engagement, taking account of local circumstances, needs and capacities; and
	o) Customising the type and level of stakeholder engagement to the needs and keeping the process flexible to adapt to changing circumstances.
Principle 11	p) **Encourage water governance frameworks that help manage trade-offs across water users, rural and urban areas, and generations**
	q) Promoting non-discriminatory participation in decision-making across people, especially vulnerable groups and people living in remote areas;
	r) Empowering local authorities and users to identify and address barriers to access quality water services and resources and promoting rural-urban co-operation including through greater partnership between water institutions and spatial planners;
	s) Promoting public debate on the risks and costs associated with too much, too little or too polluted water to raise awareness, build consensus on who pays for what, and contribute to better affordability and sustainability now and in the future; and
	t) Encouraging evidence-based assessment of the distributional consequences of water-related policies on citizens, water users and places to guide decision-making.
Principle 12	u) **Promote regular monitoring and evaluation of water policy and governance where appropriate, share the results with the public and make adjustments when needed**
	v) Promoting dedicated institutions for monitoring and evaluation that are endowed with sufficient capacity, appropriate degree of independence and resources as well as the necessary instruments;
	w) Developing reliable monitoring and reporting mechanisms to effectively guide decision-making;
	x) Assessing to what extent water policy fulfils the intended outcomes and water governance frameworks are fit for purpose; and
	y) Encouraging timely and transparent sharing of the evaluation results and adapting strategies as new information become available.

Source: OECD (2015a).

responsibilities, managing water at the appropriate scales, encouraging policy coherence, and adapting the level of capacity to the complexity of water challenges to be met.

- *Efficiency* relates to the contribution of governance to maximize the benefits of sustainable water management and welfare at the least cost to society. It relies on sharing water-related data and information, mobilizing water finance, enforcing regulatory frameworks, and promoting innovative water governance practices (inter-municipal collaboration, urban–rural partnerships, etc.).
- *Trust and engagement* relate to the contribution of governance to building public confidence and ensuring inclusiveness of stakeholders through democratic legitimacy and fairness for society at large. It is about mainstreaming integrity and

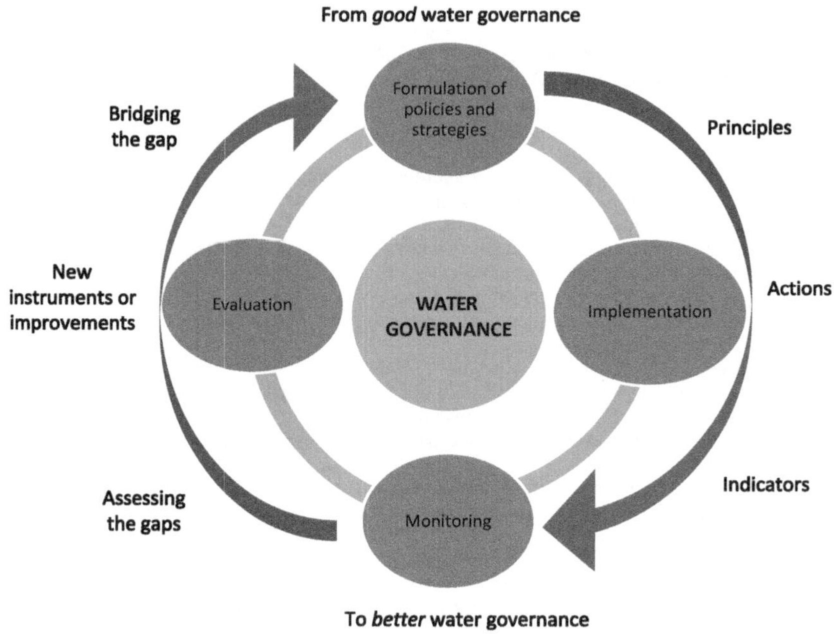

Figure 1. The Water Governance Cycle (*Source*: OECD, 2015a).

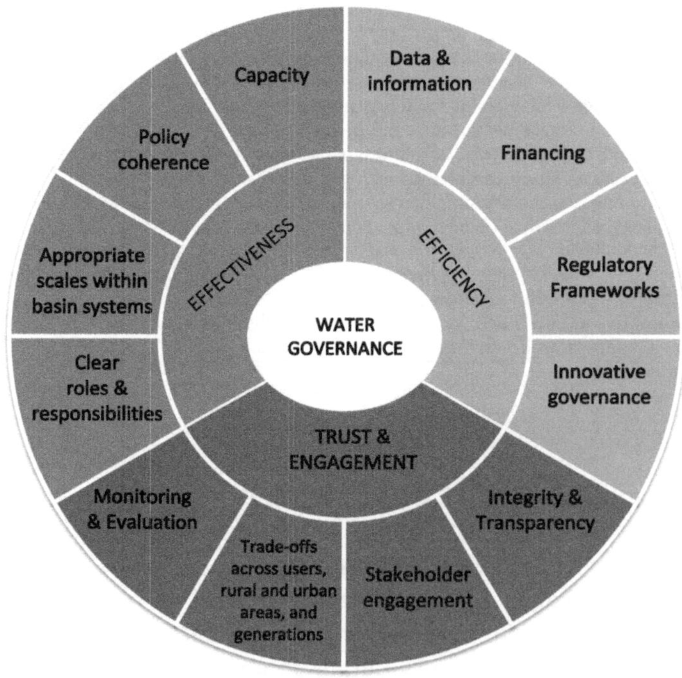

Figure 2. Overview of oecd principles on water governance (*Source*: OECD, 2015a).

transparency, managing trade-offs across water users, rural and urban areas, and generations, and promoting regular monitoring and evaluation to make adjustments when needed.

The principles are relevant for all levels of government. They apply to the overarching water policy cycle and should be implemented in a systemic and inclusive manner. As such, they apply across (1) water management *functions* (e.g. drinking water supply, sanitation, flood protection, water quality, water quantity, rainwater and stormwater); (2) water *uses* (e.g. domestic, industry, agriculture, energy and environment),; and (3) *ownership* of water management, resources and assets (e.g. public, private, or mixed).

Since their adoption by the OECD ministers, seven non-OECD countries (Brazil, China, Colombia, Morocco, Peru, Romania and South Africa) and more than 150 other actors have committed to mainstreaming the principles across their policies and activities through endorsing them formally and joining the Global Coalition for Good Water Governance set up in 2016. The coalition aims to trigger collective action and guide public action from policy makers, business, and society at large through the identification, collection and upscaling of innovative solutions.

From principles to indicators and improved policies and practices

The principles are based on the consideration that water governance is a means to an end, rather than an end in itself. Understanding the performance of governance arrangements is a means to ultimately improve the management of too much, too little and too polluted water in the short, medium and long term.

Assessing governance frameworks requires dialogue between different actors supported by robust evidence on the current state of play and needed actions to guide decision-making processes now and in the future. The OECD principles have been used by different stakeholders (practitioners, researchers, policy makers, etc.) as a framework to appraise water governance efficiency, effectiveness and inclusiveness through dialogues in a specific sub-sector, for instance flood risk governance as applied to the Dutch Flood Protection Programme (see Seijger et al. in this special issue); in a given context, such as Action Against Hunger's humanitarian and development projects on water and sanitation governance; and at a specific scale, in this case the city level as part of the *City Blueprint Approach.*

Since the adoption of the OECD principles, the OECD has been developing an implementation strategy based on the bottom-up and multi-stakeholder development of an indicator framework and the collection of water governance stories addressing some or all of the principles. The indicators are conceived as a self-assessment framework for governments and stakeholders to carry out a dialogue on their water governance systems at a given moment, to track progress over time and to map the concrete actions needed to bridge identified gaps. In particular, the indicators seek to build consensus across a range of public authorities and stakeholders on the strengths and weaknesses of water governance systems and the ways forward. Simultaneously, 60-plus water governance stories illustrating the implementation of the OECD principles have been collected at local, basin, national and global levels to showcase concrete experiences, lessons learned from success and pitfalls to avoid. Both indicators (to be published in a dedicated OECD report) and stories (to be featured in a dedicated database) are expected to encourage and facilitate the uptake of the OECD Principles on Water Governance at different levels and ultimately contribute to the development of better water policies for better lives.

Note

1. The OECD Water Governance Initiative is an international multi-stakeholder network of members from the public, private and non-for-profit sectors gathering twice a year to share good practices in support of better governance in the water sector.

Disclosure statement

No potential conflict of interest was reported by the authors.

References

OECD. (2011). *Water governance in OECD countries: A multi-level approach.* Paris: Author.

OECD. (2012a). *OECD environmental outlook to 2050: The consequences of inaction.* Paris: Author.

OECD. (2012b). *Water governance in Latin America and the Caribbean: A multi-level approach.* Paris: Author.

OECD. (2013). *Making water reform happen in Mexico.* Paris: Author.

OECD. (2014a). *Water governance in the Netherlands: Fit for the future?* Paris: Author.

OECD. (2014b). *Water governance in Jordan: Overcoming the challenges to private sector participation.* Paris: Author.

OECD. (2014c). *Water governance in Tunisia: Overcoming the challenges to private sector participation.* Paris: Author.

OECD. (2015a). *OECD principles on water governance.* Retrieved from http://www.oecd.org/governance/oecd-principles-on-water-governance.htm

OECD. (2015b). *Water resources governance in Brazil.* Paris: Author.

OECD. (2015c). *Stakeholder engagement for inclusive water governance.* Paris: Author.

OECD. (2015d). *The governance of water regulators.* Paris: Author.

OECD. (2016). *Water governance in cities.* Paris: Author.

OECD. (2017a). *Water charges in Brazil: The ways forward.* Paris: Author.

OECD. (2017b). *Enhancing water use efficiency in Korea: Policy issues and recommendations.* Paris: Author.

Aziza Akhmouch

Delphine Clavreul

Peter Glas

Addressing the policy-implementation gaps in water services: the key role of meso-institutions

Claude Ménard, Alejandro Jimenez and Hakan Tropp

ABSTRACT

This paper reviews sources of misalignment between the institutional arrangements, incentives and resources mobilized in water policies. It is argued that the resulting policy gaps develop mainly at the implementation phase and are deeply connected to flaws in the 'meso-institutions' linking both the macro-level, at which general rules are established through laws and customs, and the micro-level, at which actors operate within the domain thus delineated. It is suggested that the Organisation for Economic Co-operation and Development's (OECD) Principles on Water Governance (2015) are primarily and rightly targeting these flaws. The discussion is substantiated with numerous examples, mainly from developing countries.

Introduction

Over the past 20 years there has been significant investment in water-policy reform and supportive institutions in both developed and developing countries as means to improve the delivery of water services and promote investments. Typical reform elements are, for example, decentralization, making room for more active stakeholders and the development of independent regulatory functions (e.g., World Water Development Report (WWDR), 2006). In a developing country context, such reforms have been widely promoted and supported by bilateral and multilateral agencies. In developed countries, an increasing awareness of the consequences of ageing water and sanitation infrastructure has provided incentives for innovative solutions such as public–private partnerships.

However, notwithstanding some success, most notably in the improved access to drinking water following the achievement of Target 7c of the Millennium Development Goals (UNICEF/WHO, 2015), general access as well as the quality and sustainability of the service delivered remains a major concern for developing countries. In a developed-country context, the involvement of private operators has had mixed results, feeding debates and controversies (de Vries & Yehoue, 2013; Hall, 2015; Lobina, 2013; Ménard, 2012).

If one considers the importance of the issues at stake, with the essentiality of water for human beings, water policies are facing flaws that have not attracted the attention they deserve. Policies are hereafter understood as the set of rules, procedures and allocation mechanisms embedded in laws and regulation and that shape programmes through which services are produced and delivered (UNDP/UNICEF, 2015). Frequently, the adoption of national water policies and laws or even supranational agreements and directives has been seen as an end in itself and not necessarily the start of a longer process that requires much more than formal changes, heavy investments and commitments in financial, social and political capital and, above all, in building human capabilities. Moreover, laws and policies are not automatically implemented, administered, enforced or complied with. They require interpretation and operationalization through the development of adequate devices. Such devices, hereafter identified as 'meso-institutions', are essential to make policies and the governance of programmes associated with these policies functional, making a difference for both stakeholders and the effective management of water resources and services. In too many countries there is a substantial gap between what is stated in law and policy and what is actually happening on the ground. We identify this as the *policy-implementation gap*. Put simply, it means that the institutional arrangements, incentives and resources mobilized are misaligned, with the resulting gap that is the difference between what is stated in a law and related policies and their actual implementation.

Part of the problem comes from a misunderstanding of the nature and role of institutions involved in the definition, implementation and operationalization of these policies, and it is very fortunate that developments in the Water Governance Initiative (WGI) and their translation into the Principles on Water Governance (PWG) (OECD, 2015a) contribute to the clarification of these issues. This has been done essentially through increased attention paid to the concept of governance and to the diverse modalities through which governance operates. This focus on governance in the water sector partly intends (and helps) to overcome this implementation gap and to reorient attention onto the quality of institutions and stakeholder relationships as well as addressing power and politics as fundamental issues. Attention to modalities of governance can provide a very useful point of entry for capturing how specific policies and the laws or customary rules in which they are embedded are translated into more detailed rules, rights and procedures shaping the interactions among stakeholders.

However, 'governance' is a polymorphic term, partly because governance issues may vary according to the purpose at stake and the multiple levels of decision-making involved, from 'corporate governance' to governmental governance. This paper intends to shed a light on these issues. To avoid confusion, a distinction is made between the multiple *levels* of government through which policies are decided and implemented (e.g., supranational, national, regional, local; OECD, 2011) and the institutional *layers* in which different modalities of governance are embedded. Our analysis focuses primarily on these institutional layers, hoping to provide more robust theoretical ground to what is meant by governance and, even more importantly, to the various modalities through which governance operates. To add flesh to the framework thus developed, we rely on an assessment of major policy gaps, with special attention paid to the implementation phase, a central one in the water governance cycle (OECD, 2015a).

The paper builds on the landmark contribution by Davis and North (1971) and numerous theoreticians of institutions that followed the path they opened up (e.g.,

Aoki, 2001; Greif, 2006; Acemoglu & Robinson, 2012). In that literature, a classic distinction is between the 'institutional environment' – a notion that delineates the layer within which general rules are established, for example, laws defining and allocating users' rights over water – and the 'institutional arrangements', which refers to players, typically firms, users' associations, non-governmental organizations (NGOs), operating within the rules of the games thus defined (Davis & North, 1971). Note that the institutional environment is not necessarily defined at the 'national' level: it can be associated with supranational or regional power and, in many cases, with transboundary water entities setting the rules. Moreover, rules are not necessarily embedded in laws: in many countries, they are rooted in customary water rights.

However, there is a flaw in this framework. We shall argue that a layer is either neglected or even missing, which is about the modalities linking the macro-institutional layer at which rules are defined and the micro-institutional layer at which players operate. We identify this missing layer as that of intermediate or 'meso-institutions'. Meso-institutions designate the set of devices (entities) and mechanisms (procedures) through which the general rules are translated, adapted and made operational, providing guidelines to operators and users and feedbacks to decision-makers. Examples of devices are regulatory agencies, public bureaus, local commissions and stakeholders committees. Examples of mechanisms are administrative rules or protocols. The combination of devices and mechanisms defines different meso-institutions and makes them a central piece in which to understand and explain policy-implementation gaps. As shown below, this intermediate layer is underlying many of the principles enunciated in the PWG (OECD, 2015a).

These issues are explored and discussed primarily in relation to drinking water systems.[1] The paper is structured as follows. The next section summarizes, mainly through examples drawn from developing countries, the many faces of policy-implementation gaps that several of the PWG implicitly targeted. The third section proposes a framework to capture the different institutional layers underlying these gaps and to provide tools for better understanding their source. The fourth section relates this framework to the lessons drawn in the WGI project and suggests some possible consequences for policy-making. The fifth section concludes.

Implementation gaps: a multifaceted issue

Numerous potential gaps have been identified in the governance of water systems (OECD, 2011). The PWG (OECD, 2015a) originated in that diagnosis and provides indications for reforms that could help to overcome these gaps. Among the flaws thus pointed out, there is an increasing awareness that a substantial part proliferates at the implementation phase of the governance cycle and poses a serious challenge, especially in developing countries where providing affordable access to water and securing its sustainability remain major issues. However, the exact nature of these flaws and their causes remain underexplored. What are the main gaps and what are the reasons behind them? This section describes four groups of factors that contribute to these gaps.

Gaps in the policy-formulation process

Genuine national control and monitoring of the policy process is essential for a successful policy development. In many cases, particularly in aid-dependent countries, governments suffer from pressure (either explicit or implicit) to develop reforms that might be neither truly demanded nor appropriately embedded in the social values of stakeholders. Another gap, partly linked to external pressure, is the development of overambitious policies, setting objectives that are disconnected from current national reality and capacities. While these policies might adhere to international thinking on good practice, the goals assigned result in unrealistic targets when it comes to their implementation, as well illustrated in the analysis of the Water Sector Development Plan 2006–2025 in Tanzania (Jiménez & Pérez-Foguet, 2011a). Basic aspects of capacity and institutional arrangements for delivery were hindering a sustainable service delivery in rural areas, which showed important weaknesses in terms of functionality (Jiménez & Pérez-Foguet, 2011b), quality and year-round availability of services (Jiménez & Pérez-Foguet, 2012). The National Sanitation Campaign, developed in Tanzania in 2012, shows similar weaknesses in the design and overambitious goals compared to current capacities (Jiménez, Mtango, & Cairncross, 2014a).

Another gap that might hinder policy implementation comes from lack of high-level political commitment (Savedoff & Spiller, 1999). In many contexts the volatility of political representatives within the government involves changes in priorities. Such volatility is particularly damaging in the water sector because of the amplitude of sunk costs and the long-term investments required, so that this sector faces a particularly challenging gap between the political and business cycles when it comes to the implementation of reform. Securing long-term political commitment is easier said than done, but it is essential in services such as water, which require long-term dedicated investments.

Building institutional devices that allow the inclusion of subnational governments, private and community stakeholders, making room for 'voice' in the policy-making process, is a necessary condition to craft realistic, implementable policies (Hirschman, 1970; OECD, 2011, 2015a). Particularly important in that respect is the capacity to tailor the participatory processes in a way that mitigates power imbalances between stakeholders and within communities (COHRE, AAAS, SDC & UN-Habitat, 2007), an inclusive strategy of particular significance when ethnic minorities and indigenous peoples are concerned (Jackson, Tan, Mooney, Hoverman, & White, 2012; Jiménez, Cortobius, & Kjellen, 2014b).

A partially related factor is the risk of policy capture by economic or political elites. Elite capture means that a few powerful and resourceful groups or individuals use their influence to usurp resources, skewing policy formulation or implementation (Shirley, 2002, 2008). Decentralization in the water sector has, for example, been criticized for failing to grasp properly social and political dynamics at the local level, which opens up the possibility of elite capture in local water users' associations. For instance, there are many examples in developing countries where publicly funded bore-wells in rural areas are ending up in the backyard of a local strongman, preventing other villagers from making use of the well or obliging them to pay for rights of access (Rusca, Schwartz, Hadzovic, & Ahlers, 2015). Elite capture can also take place further upstream in the policy cycle. For example, water markets in Chile were captured by powerful buyers hoarding water rights that worked against more equitable water distributions to local farmers (Warner et al., 2009). Well-intentioned policy measures such as decentralization, public–private partnerships or market-based water allocations without the proper checks and balances and without adequate administrative capabilities (OECD, 2015a) may lead to local elites strengthening their positions at the expense of politically and economically marginalized groups.

Gaps in the operationalization of the policy

Once approved, policies need to be operationalized. Many obstacles can show up at this stage. In many occasions, the support of external agencies is focused on the policy-development phase, and the governments find themselves with approved policies that are not completely understood and with limited support, or even no support at all, for their implementation. In many cases, policy-making is ascribed to decision-makers, while implementation is linked to administrative capacity (UNESCO, 2006). One very common problem is the mismatch between the responsibilities allocated to certain 'meso-institutions' (local governments, regulatory agencies etc.), and the resources that flow into them (Crook, 2003; Jiménez & Pérez-Foguet, 2011c; OECD, 2015a; Ribot, Agrawal, & Larson, 2006). Beside the need for adequate funding, the human and technical capacity required cannot be acquired overnight. Policies establishing new roles for certain institutions or transferring new decision rights to devices in charge of implementing a policy need to deal with human resources that often takes time to train and adapt.

Moreover, the successful operationalization of policies depends on their perception as legitimate, which requires them to be clearly understood as well as effectively disseminated among all relevant stakeholders (OECD, 2015a; Stockholm Environment Institute (SEI), 2013). This requires a concerted effort to inform and train these stakeholders, particularly at the local level, which is often not planned for and might involve unexpected economic as well as political costs. The experience of users' commissions in France (Commission Consultative des Services Publics Locaux) or in England and Wales (Consumer Council for Water) illustrates the difficulties to be faced, even in countries with relatively well-informed stakeholders.

A particular challenge impeding implementation is the misalignment between government-induced water policies and informal water institutions such as customary water rights. Failing to understand the role and functions of deeply embedded informal institutions and how they can be best included in the policy threatens its feasibility. For example, in the Central Asian former Soviet Union countries of Tajikistan, Kyrgyzstan and Uzbekistan it has been observed that there are competing institutional regimes based on recent integrated water resources management (IWRM) reform, an inherited Soviet managerial style of command and control, and customary water rights. Local actors typically combine newly established rules of local water management (water users' associations) with informal institutions that often originate from institutionalized Soviet and pre-Soviet patterns of behaviour (Schlüter & Herrfahrdt-Pahle, 2011; Sehring, 2009). This is a recipe made for confusion of which rules and regulations should apply, fuelling discretionary decisions in the operationalization of well-intentioned policies.

Path dependence and resulting misalignments can thus lead to formal institutions disrupted or completely replaced by informal ones, or to well-functioning informal institutions that are undermined. While some informal institutions can be well functioning and support reform aspirations, many are rather fuelling discretionary and clientelistic decision-making seriously distorting implementation. A case from Paraguay shows how better alignment between formal and informal institutions can benefit both sides. In Paraguay, informal private water-supply systems were recognized and agreements developed between local government entities and small-scale private water vendors. The outcome was easier control and monitoring of pricing and quality of service (Phumpiu & Gustafsson, 2009). Similar relatively successful combinations of formal and informal arrangements have been observed in other countries (e.g., Mozambique, Blanc, 2009; or Cambodia, Frenoux, 2016).

The underlying lesson is that successful policy implementation requires cooperative relationships between many parties operating within different institutional layers: government at various levels, the private sector and community-based organizations. When appropriate incentives to cooperate are not built into the policy framework, it is unlikely that actors that might have limited trust to each other will provide adequate support to policy implementation. Because of the centrality of water to human survival, suppliers, whether private or public, are often perceived as having disproportionate power, a perception often reinforced by information asymmetry among parties. Without institutions that have sufficient capacity to monitor and enforce agreed norms, and/or if users do not have adequate channels to signal their requests or express their dissatisfaction, incentives to implement welcome policies will likely be weakened or derailed, even paralyzed (OECD, 2015a).

Gaps related to the characteristics and behaviour of stakeholders

A part of these implementation problems is rooted at the micro-level, coming out of the variety of stakeholders involved in the provision as well as the usage of water, with actors that have different, often diverging or even conflicting, interests. Having appropriate institutional devices to monitor their rights and adequate institutional decision-making mechanism to coordinate their interests and eventually arbitrate among them are key factors in making a policy accepted or rejected.

On the supply side, an important characteristic of drinking water is the frequent monopolistic position of providers, at least in cities of a significant size,[2] which raises the classic problem of this market structure with respect to the delivery of adequate quantity at an appropriate price (OECD, 2011). The acceptance of institutional solutions to deal with these risks is a tributary of the type of regulatory arrangements and on the capacity of regulatory devices to benefit from adequate human and financial resources, to be able to constrain deviants and to resist capture by operators (OECD, 2015a).

However, the risk of capture is not solely on the supply side. Spiller (2009) has shown that it can also come from what he calls ‘third party opportunism’, a situation in which interest groups that are particularly influential or which can build an active coalition may create powerful obstacles to the implementation of legitimate policies (e.g., preventing taxation of the benefactors of swimming pools in an environment with a scarce water resource) or distort policies through their influence on the decision-making process. The creation of representative organizations of consumers (e.g., water councils) does not automatically solve the problem since it depends so much on the quality of the representatives of stakeholders and the individual as well as institutional capacity to avoid capture by specific interest groups. This risk of third-party opportunism also raises the sensitive issue of possible conflicts between so-called 'representative' organizations and the responsibilities of elected decision-makers. Low participation of water users and the poor quality of those claiming to be their representatives, which is often fed by lack of resources and programmes to support their role, can therefore create gaps that they were intended to fill!

These different sources of opportunistic behaviour often end up in corruption, another and often overlooked source of the policy-implementation gap (e.g., Transparency International, 2008; Water Integrity Network, 2015). In many contexts corrupt behaviour has become the norm, fuelled by political patronage. Ironically, corruption can constitute an institution in its own right, following its own rules and regulations.

Corruption takes many forms, ranging from petty to grand or political corruption. Examples include:

- collusion between water consumers and local civil servants to rig water meters to avoid paying water fees, or consumers being pressurized to pay local civil servants for receiving the water service in the first place;
- bribery related to the awarding of licences for water uses and wastewater discharges;
- collusion (kickbacks or bid-rigging) and extortion in procurement procedures and the awarding of contracts;
- nepotism and kickbacks in the appointment and promotion to higher-level civil servant positions;
- collusion during the quality control of the construction of water infrastructure works; and
- embezzlement of government and foreign aid funds and assets

Endemic corruption undermines any water policy as well as administrative rules and regulations. It also erodes critical foundations of trust, the rule of law, fairness and efficiency of water institutions. Indeed, corruption and 'red tapism', often overlapping with bureaucratic inertia, increase transaction costs, discourage investments and feed strong disincentives for water-reform implementation. At its core corruption is a strong symptom of institutional weaknesses and poor governance. Recent studies confirm that even petty corruption at the water provider–consumer interface is one key risk that water sector must face (Butterworth & de la Harpe, 2009). This calls for more effective enforcement of regulatory institutions to monitor performance and expenditures of service providers (OECD, 2015a).

Gaps related to the overarching country-governance situation

Water policies do not exist in a vacuum; they depend on the overarching governance system (UNDP/SIWI, 2015). Several studies, going back at least to Shirley (2002), suggest that changes in water governance are connected to other governance reforms.[3]

Indeed, water policies might be hindered by more general institutional factors that inhibit tentative reforms of governance (Jiménez, Jawara, LeDeunff, Naylor, & Scharp, 2017). For example, the establishment of regulatory body for water services in Bosnia and Herzegovina faced several obstacles derived from the complex administrative and political setup that prevented its implementation. Similarly, recent research shows how lack of accountability and poor top-down discipline within the public sector are among the major hindrances for effective service delivery in sub-Saharan Africa, including water (de Mariz, Ménard, & Abeillé, 2014; Overseas Development Institute (ODI), 2015). Fragile and post-conflict countries, which have among the lowest levels of coverage of water and sanitation services, face particular challenges. In most cases, the state has not the basic capacities to steer the sector, while the extension of service is heavily driven by external support, not always carried out in a very coordinated manner. High pressure to deliver on many different development priorities as well as the vulnerability to external shocks makes it even more complicated. For example, in Liberia, whose awful civil war ended in 2003, it was found in 2011 that over 95% of water points in the rural areas had been developed in the last decade by external agents, without coordination and control over conditions of delivery of water (Government of Liberia (GoL), 2011a), which in turns undermines the government's capacity to ensure sustainability of the services. The governance of the sector is fragmented across various ministries, and government commitment to reform the sector, under the Liberia WASH Compact of 2011 (GoL, 2011b), remains to be implemented to a large extent.

The lesson from such cases is complex. On the one hand, delivering on policies to improve services can pay a double dividend in fragile states of enhancing well-being of citizens as well as gain state legitimacy (Brinkerhoff, Wetterberg, & Dunn, 2012). On the other hand, lack of adequate democratic culture, including debate, consultation and participation, represents a major challenge to the legitimacy of the policies adopted. More generally, improving governance in the water sector cannot be dissociated from the development of institutional devices through which stakeholders can be appropriately informed and can provide feedback to policy-makers. Building such institutions is a crucial element to effective reform, facilitating the implementation of policies chosen and reducing the economic and political transaction costs of that process.

Policy gaps: institutions matter … under different modalities

The policy gaps pointed out above and summarized in Table 1 are unambiguously embedded into institutional flaws and/or the dynamics of actors operating within mismatched rules and norms. However, 'institutions' are not an undifferentiated magma. They rather define a complex setting within which policies are elaborated, implemented … and often distorted! The following sections describe the concept of institutional layers and how they relate to geographical and administrative levels.

Institutional layers: a much needed clarification

Indeed, much confusion remains when it comes to the specific components grouped under the terms 'institutions' and 'governance'. The polymorphic meaning of 'governance' illustrates the difficulty: the term refers to very heterogeneous problems and serves

Table 1. Type of gaps in policy implementation, and typical causes.

Policy-implementation gaps	Causes
Gaps in the policy-formulation process	Lack of national oversight over policy formulation External pressure to adopt blue-print policies not adapted to the context Lack of high-level political commitment Lack of participation in policy formulation Policy capture by elites or influential groups
Gaps in the operationalization of the policy	Mismatch between responsibilities and resources Time needed to build capacity is not adequately considered Lack of legitimacy of institutions that implement the policy Misalignment between water policies and informal water institutions Lack of capacity to monitor and enforce the agreed norms No channels to signal users' demands or express dissatisfaction
Gaps related to the characteristics and behaviour of stakeholders	Monopolistic position of providers 'Third-party opportunism' Quality of the representation of stakeholders Capture of stakeholder representation by specific interest groups Corruption
Gaps related to the overarching country-governance situation	Political instability, protracted crisis and insecurity Government's lack of capacity to conduct basic functions Lack of accountability in the public sector Poor top-down discipline in government No practice of democratic culture, including debate, consultation and participation

various purposes (Tropp, 2007). For example, some contributions understand governance as the procedures needed to monitor rules of the game defined at the macro-level, such as the judicial procedures required to implement laws (World Development Report (WDR), 2017). In management, governance often refers to 'corporate governance', that is, modalities to pilot firms or other micro-institutions (NGOs, international organizations etc.). Still, other contributions interpret governance in the context of devices designed to implement policies understood as a set of rules, procedures and allocation mechanisms embedded in laws and regulation (OECD, 2015b; UNDP Water Governance Facility/UNICEF, 2015). These ambiguities are regrettable if one considers the importance of the issues as stake, which all relate to the implementation of rules. For example, recent OECD reports (e.g., OECD, 2009, 2015b, 2017) and numerous specific contributions to the WGI (e.g., the case studies on the Netherlands, OECD, 2014; or Brazil, OECD, 2015d), made a big step forward by taking on board the role of institutions in the development of policies and the explanation of their flaws. But they also suffer from these ambiguities, largely due to the mess in the representation of the different institutional layers in which 'governance' is embedded.

Building on Davis and North (1971), Williamson (1985), Aoki (2001), Ostrom (2005), Greif (2006), among many others, recent developments suggest disentangling these institutional layers (Hodgson, 2015; Kunneke, Groenewegen, & Ménard, 2010; Ménard, 2017). In a sense, reports issued in the context of the WGI participate to this evolution through their effort to differentiate levels of government within which policies are elaborated and their governance implemented (OECD, 2011, 2015a, 2016). However, 'levels of government' is a limited unit of analysis when it comes to understanding policy gaps identified in the previous section. What matters most are the different institutional layers, of which 'government levels' are only a subset, within which these gaps are embedded.

Our starting point finds inspiration in the Northian perspective on *institutions* understood as 'the set of fundamental political, social, and legal ground rules that establishes the basis for production, exchange, and distribution' (Davis & North, 1971, p. 6) or, more succinctly, the 'integrated systems of rules that structure social interactions' (Hodgson, 2015, p. 57). 'Institutions' thus defined is therefore an overarching concept that captures different layers through which rules and norms are defined, implemented and operationalized.

In that respect, three layers can be distinguished. At the most general level are *macro-institutions*, within which constitutive rules and norms are defined and rights and duties established, delineating the domain of possible actions. For instance, the adoption of a law opening room for public–private partnership in water utilities typically belongs to this macro-institutional layer, while the entities created or restructured in that new legal context can be identified as micro-institutions (see below). Macro-institutions are 'configurational' (Ostrom, 2014): they determine who can hold what rights with what responsibilities, what can be done, and the constraints under which it can be done. In democratic regimes this is typically the role of the political system (e.g., a parliament) and/or the judiciary. Many policy gaps identified above in the second section come from flaws in these macro-institutions: poorly designed rules of the game such as badly defined property rights can introduce loopholes in procedures of implementation, opening room for opportunistic behaviour all the way up to bribery and corruption. Note that macro-institutions cannot be restricted to the sole formal, legalistic rules. North (1990a) introduced an important distinction between formal and informal (macro-)institutions, a distinction relevant for the analysis of the water sector as emphasized in a UNDP report:

> Formal, or statutory, institutions exist at many different levels and can have a direct and indirect impact on water. A clear example of a formal institution is a national constitution, which provides the framework for all other legislation and rules and regulations in a given country. In South Africa, for example, the right to water was enacted in the constitution to redress past racial discrimination. [...]
>
> Informal water institutions refer to traditional and contemporary social rules and norms that decide on water management, use and allocation. [...] Large shares of countries' water resources are allocated on the basis of customary water rights. Small-scale farming is still a main occupation in many developing countries, and a large share of the water resources being used in irrigation is largely outside the regulatory control of the government. This does not necessarily mean that water resources are unregulated, since farmers may agree among themselves on what rights and obligations should apply for water use and management. Nor does it mean that informal water rights systems are 'archaic'. On the contrary, they can comprise a dynamic mix of principles and organizational forms of different origins.(UNDP, 2013, pp. 9–10)

Social rules and norms matter because they shape beliefs and expectations that structure the behaviour of actors. Indeed, it is within these rules and norms that actors organize transactions needed for the production and delivery of water services, thus delineating the *micro-institutional layer*. This layer is the domain of water operators, water cooperatives, NGOs involved in the delivery of water and various users, actors that modern economic theory captures through the generic concept of 'organizations' (Gibbons & Roberts, 2013; Ménard, 2008; Milgrom & Roberts, 1992; Williamson, 1985, 1996).

Meso-institutions and the governance of water systems

However, there is a missing link in this picture, a gap between the layer at which general rules are defined and the layer at which actors make their decision in the context thus delineated. Indeed, macro-institutional rules and norms are rarely self-fulfilling and strictly implemented by micro-institutional actors. Rules and norms need to be 'interpreted, 'translated' and 'monitored'. For example, laws intending to guarantee more transparency in public procurement and to reinforce accountability might be adopted by a well-intentioned parliament; they still require devices to be implemented and monitored (de Mariz et al., 2014). Propositions regarding 'The governance of water regulators' (OECD, 2015b), or the need for a multilevel approach to governance in the water sector (OECD, 2011), typically relate to this issue. However, although such contributions rightly emphasize that there are different levels of government involved, they do not capture the role of devices often transversal to governmental levels and that play a central role in the implementation of policies. These devices fall into the category of what can be called intermediate or *meso-institutions*: this is the layer in which general rules and rights are interpreted, implemented, monitored and controlled. Regulatory agencies are such devices. Other examples are ministerial departments in charge of the water sector (typically a department of public works), local authorities responsible for the organization of the sector when water management is decentralized, self-organized local communities monitoring the resource etc. (Marques, Simões, Pires, & Almeida, 2009; OECD, 2015b).

Many policy gaps identified in the second section come from flaws in the design of these meso-institutions or in responsibilities they are allocated. Among the challenges they face is their capacity to operate both ways when bridging the gap between macro- and micro-institutions, facilitating the implementation of policies (top down) and channelling information about the social demand of users (bottom up). Another challenge concerns their capacity to adapt the general rules to specific situations in time, space and scope. Monitoring operator(s) in a city of several million inhabitants substantially differs from accomplishing the same task in a remote village, notwithstanding that both situations may fall under the same general laws. Similarly, general rules regarding tariffs may be misaligned with hydrological characteristics or users' income of a specific region, thus doomed to fail in the absence of appropriate meso-institutions to adapt. Setbacks of many well-intentioned programmes of international organizations and development agencies are illustrative of failures to build intermediate devices that can adjust general rules to such context-dependent factors, as so well illustrated by the example of irrigation programmes in Nepal (Ostrom, 2014).

Most analytical approaches to meso-institutions have so far focused on regulatory agencies and the conditions of their efficiency, e.g., their independence from policy-makers, the availability of adequate financial resources etc. (Armstrong, Cowan, & Vickers, 1994; Laffont, 2005; Laffont & Tirole, 1993; OECD, 2015b; Spiller, 2009). On the empirical side, the diversity of institutional devices operating as meso-institutions is better acknowledged (OECD, 2011, 2015a, 2015b, 2016; Tremolet & Binder, 2010;

Table 2. Interaction between the institutional layers and the administrative and geographical levels.

Administrative and geographical levels	Institutional layers		
	Macro-institutional layer (policy-making)	Meso-institutional layer (interpretation and oversight)	Micro-institutional layer (delivery/ implementation)
Supranational	United Nations, European Union, regional economic commissions etc.		
National	Parliament/judiciary system	Regulators, ministerial departments in charge of water	
Subnational region	Regional parliament	Regulator (subnational)	
District/municipal		Water municipal unit	Water utilities, corporate water users
Local/village	Customary water rights	Village monitoring committees	Small service providers; individual water users

UNDP, 2013). Efforts to develop a typology of these diverse arrangements can be found in Marques (2010), the OECD report on water regulators (OECD, 2015b), and Ménard (2017), among others. Table 2 summarizes (and illustrates) these institutional layers and their positioning at different administrative/geographical levels.

Building credibility to address policy gaps: political commitment and political transaction costs

The prevailing focus on regulatory agencies and the presumption of their superior efficiency tends to ignore factors that can make this solution fragile, threatening the implementation of water policies and the capacity to make them operational (see the second section).

First, one of the greatest challenges to efficient meso-institutional devices, particularly in developing as well as in 'emerging' economies, is the availability of competent human resources (OECD, 2015a, Principle No. 4). To be perceived as adequate and legitimate, staffing must be provided through clear standards with respect to: (1) entry-level qualification; (2) modalities of selection; (3) the appropriateness of remuneration; (4) the provision of a career path; and (5) the dissemination and enforcement of these standards (de Mariz et al., 2014). These conditions remain relevant whatever the type of meso-institutions at stake (regulatory agency, public bureau, local council etc.). Too many 'light' training programmes do not address this need to build human capital or, even worst, do not take into account the actual conditions under which staff operate, opening room to incompetent decisions ... or corruption (see the second section; and OECD, 2015a, Principle Nos 4 and 9).

Second, water policies take time to be defined and implemented, which can undermine the credibility of policies and policy-makers. Indeed, a general policy problem, particularly acute in democratic regimes, is that political cycles do not match business cycles, making it hard for even well-intentioned policy-makers to adopt and stick to policies not matching their electoral cycle. In water systems, this lag between cycles is amplified by the long period required for the appropriate management of the resource and for recovering the sunk costs of the highly dedicated investments it needs (Savedoff

& Spiller, 1999). The resulting difficulty of making commitments credible makes even more central the role of meso-institutions in the effective implementation of the policies adopted.

Third, building such meso-institutions involves significant transaction costs. *Economic transaction costs* are the costs of planning, monitoring and bringing to its achievement the allocation and transfer of rights to use the resource or its associated services (Williamson, 1996). Examples are the costs of contracting with a private operator, or monitoring the delegation of management to local authorities. Most analyses of these costs have so far focused on the micro-institutional level, e.g., the costs and benefits of operating through contracts between a public authority and a private operator rather than offering the service through a public monopoly.[4] However, there are also costs attached to meso-institutions, such as those of running a regulatory agency, the bureaucratic costs of a public department monitoring water services etc. These costs may well challenge the appropriateness of the solution selected to implement water policies (for illustrations, see Ostrom, 2005, 2014). Unfortunately, there are very few such comparative assessments so far (but see Libecap, 2014, and Vatn, 2015, for environmental policies; and Ménard, 2016, for fisheries).

Fourth, many policy gaps identified above are coming out of what North (1990b) called *political transaction costs* (see also Marshall & Weingast, 1988; Robinson, 2010). With respect to meso-institutions, political transaction costs are the costs of building and monitoring coalitions among actors in charge of implementing the rules and norms that organize the sector. For instance, once a legal framework has been adopted that favours a public–private partnerships to develop or modernize water infrastructures, there are costs to build and implement an agreement between, say, local authorities, a private operator and representative of users (e.g., consumers' councils). These costs might create gaps between what is expected from the policy adopted and the conditions of its implementation. They might even be prohibitive, as when the behaviour of some parties creates rigidities that disrupt the development of a project, what Spiller (2009) identified as 'third party opportunism' and risk of capture by specific groups of interests (see above).

Unfortunately, there are so far very few empirical studies assessing these costs associated with the existence and running of meso-institutions (but see Grafton, Libecap, Edwards, O'Brien, & Landry, 2012). However, numerous indicators have recently been proposed in the water sector to identify these costs and measure them (see the pioneering contribution by Araral & Yu, 2013; Ménard, 2016; OECD, 2015c, 2016). Developing these tools should be high on the agenda of researchers and organizations working on policy gaps in water services (as well as in other 'essential services').

Towards a better implementation of water policies: revisiting the PWG

The analysis developed above has identified gaps that pertain to policy formulation, policy operationalization, the characteristics and behaviour of stakeholders, and to the overarching institutional environment in which these gaps are embedded. It has also emphasized the importance to analyze and strengthen meso-institutions for more effective policy implementation, opening room to a more focused interpretation of governance issues.

The PWG (OECD, 2015a) represents an important step in the direction to overcome the gaps thus identified and deal with the relevant institutional factors. Indeed, notwithstanding the significant investments made, particularly following the impulse of the Millennium Development Goals and the efforts to understand better the obstacles to water reforms and how they can be overcome, the questions remain: How can the prospects for water policy implementation be improved? What institutions can support and facilitate this implementation? As noted by Cordoba (1994), reform will be successful if there is economic rationale in its design, political sensitivity in its implementation, and close and constant attention to political–economic interactions between social and institutional factors. In particular, there needs to be a more complete understanding of the forces that lead to policy development in the first place and, critically, a concerted drive to make sure that policies are followed through to implementation. There also needs to be effective feedback and assessment mechanisms, so that the consequences of policy implementation can inform future policy development.

An innovative and encouraging aspect of the WGI and the PWG that came out of its activities is that they clearly put issues of institutions and governance at the core of the adequate definition and implementation of policies in water systems. The 12 principles, elaborated through a multi-stakeholder process, synthesize a substantial part of the work done by the WGI in the previous years and encapsulate many of the critical governance elements that need to be addressed to strengthen water policies. However, countries need to assess their own policy-implementation gaps, set priorities on how to move forward to close some of them, and establish adequate institutional devices to reach these goals. A main feature of the PWG is that they are the only internationally recognized set of guiding recommendation on water governance, which as such can add political weight. The PWG do not tell countries what priorities they should undertake or how they should do it; their merit rather lies in that different clusters of principles can provide countries a basis on which they can assess and set priorities for how to improve water governance with the aim to strengthen vital functions of particular water institutions and organizations (OECD, 2011, 2015a, 2015b).

Table 3 suggests a framework that links policy gaps to particular governance responses as outlined by the PWG. It provides a frame on which the role of meso-institutions can be analyzed and improved up on.

For example, Principle No. 7 addresses directly gaps identified above (on institutions required for making policies operational; and on gaps related to the overarching country-governance situation). Indeed, it emphasizes the need for well-functioning meso-institutions and states the need for ensuring 'that sound water management regulatory frameworks are effectively implemented and enforced in pursuit of the public interest'. It stresses the importance of coherent and predictable outcomes of administrative rules, standards and guidelines. It also points to issues of participation, coordination, transparency, and rewards and penalties to promote decision-making outcomes that are non-discriminatory, cost-effective and long-term. It draws attention to the need to endow regulators with both financial and human resources (see also Principle No. 4), a point emphasized above as central to effective meso-institutions. Another example is Principle No. 9 that points out some fundamental conditions (particularly integrity and transparency) that policies must meet to be credible, which requires a clear allocation of role and responsibilities (Principle No. 1) and modalities of

Table 3. Policy-implementation gaps in relation to the Organisation for Economic Co-operation and Development's (OECD) Principles on Water Governance (PWG) (2015a) (and underlying meso-institutions).

Policy-implementation gaps	Link to the PWG
Gaps in the policy-formulation process	PWG 1: Clearly allocate and distinguish roles and responsibilities for water policy-making, policy implementation, operational management and regulation, and foster coordination across these responsible authorities PWG 4: Adapt the level of capacity of the responsible authorities to the complexity of the water challenges to be met, and to the set of competencies required to carry out their duties PWG 10: Promote stakeholders' engagement for informed and outcome-oriented contributions to water policy design and implementation
Gaps in the operationalization of the policy	PWG 1: Institutional roles and responsibilities (see above) PWG 4: Capacity (see above) PWG 6: Ensure that governance arrangements help mobilize water finance and allocate financial resources in an efficient, transparent and timely manner PWG 7: Ensure that sound water-management regulatory frameworks are effectively implemented and enforced in pursuit of the public interest PWG 9: Mainstream integrity and transparency practices across water policies, water institutions and water-governance frameworks for greater accountability and trust in decision-making PWG 12: Promote regular monitoring and evaluation of the water policy and governance, where appropriate, and share the results with the public and make adjustments when needed
Gaps related to the characteristics and behaviour of stakeholders	PWG 9: Integrity and transparency (see above) PWG 10: Stakeholder engagement (see above) PWG 4: Capacity (see above)
Gaps related to the overarching country-governance situation	PWG 7: Regulation (see above) PWG 3: Encourage policy coherence through effective cross-sectoral coordination, especially between policies for water and the environment, health, energy, agriculture, industry, spatial planning and land use

governance that a multi-stakeholders approach can help to implement and evaluate (Principles Nos 10 and 12).

As these examples suggest, the implementation of the PWG requires appropriate institutional design (see the third section of this paper). It is even more so since these principles are mutually supportive. They reinforce each other, which also means that institutional flaws weakening one principle can have a negative spillover effect, thus feeding policy gaps. For example, the implementation of Principle No. 7 on regulatory frameworks also hinges on realizing clear division of roles and responsibilities for policy-making (our macro-institutional layer), implementation (the meso-layer) and operational management (the micro-layer) and their effective coordination (Principle No. 1). Poor institutional design with blurred responsibilities undermines the accountability of policy-makers and decision-makers in charge of implementing the rules of the game. Similarly, the absence of stakeholders in the decision-making process or poor representation of key subgroups (e.g., minorities, low-income users) may challenge inclusiveness and trust in regulatory decisions, leading to capture by local elites and forms of corruption (see the second section of this paper) that makes decision-making unpredictable and driven by other aims than satisfying consumers and citizens and contribute to the greater public good.

To be able to achieve upward and downward accountability and to reach non-discriminatory outcomes, meso-institutions need to be designed (or reformed) in a way that improves integrity, transparency and participation (Principle No. 9). Implementing reliable institutional procedures to monitor and evaluate policies adopted (Principle No. 12) is essential to build legitimacy for the policies chosen and to avoid increasing gaps that can undermine efforts to provide universal access to safe drinkable water at sustainable economic and political transaction costs. It is also key to the preservation of the resource, an aspect that may not be emphasized enough in the existing version of the PWG.

Conclusions

This paper has focused on the different sources of policy-implementation gaps. From a practical point of view it is not a small operation to fill these gaps. The operationalization of water services policies and laws requires an appropriate institutional environment to develop and set in place a number of rules, standards and procedures for regulating water utilities.

To reach this goal and implement adequate modalities of governance, there is a need to understand the different institutional layers involved and how they relate to administrative and geographical boundaries. Our analysis emphasized the particular role of the meso-institutional layer in that respect. We argued that meso-institutions provide an essential link between the macro-level at which rules framing activities of water services are decided by policy-makers and/or the judiciary and the micro-level at which operators implement these rules through the actual delivery of services. Indeed, it is through this intermediate layer that rules are interpreted, translated into guidelines, procedures and protocols, and monitored by various institutional devices (public bureaus, autonomous regulators, local authorities etc.).

In that respect, we suggested that the principles developed in the context of the WGI make an important step in addressing the policy gaps identified by taking on board the institutional settings and the diverse modalities of governance they command. It is highly fortunate that in doing so the PWG, rather than providing a general toolkit unfit for most situations, delivered and widely diffused to policy-makers recommendations that they must keep in mind when considering policy changes or when simply reviewing conditions to meet for a successful implementation of the existing policies.

Notes

1. Although many issues discussed below are relevant for other aspects of water management (e.g., irrigation), drinking water raises specific problems (criticality for human survival, absence of a substitute, monopolistic characteristics of its delivery, and important externalities with respect to health and the environment).
2. The situation is different in the countryside, at least when there is the possibility of digging wells, which raises different problems (mainly related to the sustainability and quality of the resource).
3. Other landmark references for different sectors are provided by Armstrong et al. (1994) and Levy and Spiller (1996). See also Tomson (2009) for a broader perspective.

4. For surveys of the transactions costs of alternative regulatory arrangements, see Ménard and Ghertman (2009) and Libecap (2014).

Acknowledgements

The authors benefited from very helpful comments and feedback from Aziza Akhmouch, Delphine Clavreul, the editorial staff of this journal and two anonymous referees. The paper also owes a lot to interactions over the years with participants to the Water Governance Initiative (OECD). The usual disclaimer fully applies.

Disclosure statement

No potential conflict of interest was reported by the authors.

ORCID

Claude Ménard http://orcid.org/0000-0002-3584-0232

References

Acemoglu, D., & Robinson, J. (2012). *Why nations fail. The origins of power, prosperity, and poverty*. New York: Crown Business.

Aoki, M. (2001). *Toward a comparative institutional analysis*. Cambridge, MA: MIT Press.

Araral, E., & Yu, D. (2013). Comparative water law, policies and administration in Asia: Evidence from 17 countries. *Water Resources Research*, *49*, 5307–5316. doi:10.1002/wrcr.20414

Armstrong, M., Cowan, S., & Vickers, J. (1994). *Regulatory reform – Economic analysis and British experience*. Cambridge, MA: MIT Press.

Blanc, A. (2009, August). *Les petits opérateurs privés de la distribution d'eau à Maputo: D'un problème à une solution?* [Small private operators in the distribution of water in Maputo: From a problem to a solution?]. Documents # 85. Paris: Agence Française de Développement.

Brinkerhoff, D. W., Wetterberg, A., & Dunn, S. (2012). Service delivery and legitimacy in fragile and conflict-affected states: Evidence from water services in Iraq. *Public Management Review*, *14*(2), 273–293. doi:10.1080/14719037.2012.657958

Butterworth, J., & de la Harpe, J. (2009). *Not so petty: Corruption risks in payment and licensing systems of water*. Brief No. 26. Bergen, U4 Anti-Corruption Resource Centre.

COHRE, AAAS, SDC and UN-Habitat. (2007). *Manual on the right to water and sanitation*. Retrieved from http://www.worldwatercouncil.org/fileadmin/wwc/Programs/Right_to_Water/Pdf_doct/RTWP__20Manual_RTWS_Final.pdf

Cordoba, J. (1994). Mexico. In J. Williamson (Ed.), *The political economy of policy reform*. Washington, DC: Institute for International Economics.

Crook, R. (2003). Decentralisation and poverty reduction in Africa: The politics of local–central relations. *Public Administration Development*, *23*, 77–88. doi:10.1002/(ISSN)1099-162X

Davis, L. E., & North, D. C. (1971). *Institutional change and American economic growth*. Cambridge: Cambridge University Press.

de Vries, P., & Yehoue, E. (2013). *The Routledge companion to public–private partnerships*. London: Routledge.

Frenoux, C. (2016). *Institutions et transactions: Déterminants et performances des services non conventionnels d'approvisionnement en eau dans les villes en développement Le cas des entrepreneurs privés locaux dans les petits centres urbains du Cambodge* [Institutions and transactions: Determinants and performance of non-conventional water services in developing cities.

The case of private local entrepreneurs in small urban centers of Cambodia] (PhD thesis). Université de Toulouse.

Gibbons, R., & Roberts, J. (Eds.). (2013). *The handbook of organizational economics*. Princeton: Princeton University Press.

Government of Liberia. (2011a). *Liberia water point atlas 2011*. Retrieved from http://wash-liberia.org/wp-content/blogs.dir/6/files/sites/6/2013/01/Final_Review_Version_-_Waterpoint_Atlas___Investment_Plan_x1.pdf

Government of Liberia. (2011b). *Review of the implementation of Liberia WASH compact*. Retrieved from http://wash-liberia.org/wp-content/blogs.dir/6/files/sites/6/2013/01/Review_of_the_Compact_-_final_version.pdf

Grafton, R. Q., Libecap, G., Edwards, E., O'Brien, R. J.., & Landry, C. (2012). Comparative assessment of water markets: Insights from the Murray–Darling basin of Australia and the Western USA. *Water Policy*, *14*, 175–193. doi:10.2166/wp.2011.016

Greif, A. (2006). *Institutions and the path to the modern economy: Lessons from medieval trade*. New York: Cambridge University Press.

Hall, D. (2015, February). *Why public–private partnerships don't work. The many advantages of public alternative*. London: PSIRU and OPSI, University of Greenwich. Last consulted: January 2017.

Hirschman, A. O. (1970). *Exit, voice, and loyalty*. Cambridge, MA: Harvard University Press.

Hodgson, G. (2015). *Conceptualizing capitalism. Institutions, evolution, future*. Chicago: University of Chicago Press.

Jackson, S., Tan, P. L., Mooney, C., Hoverman, S., & White, I. (2012). Principles and guidelines for good practice in Indigenous engagement in water planning. *Journal of Hydrology*, *474*, 57–65. doi:10.1016/j.jhydrol.2011.12.015

Jiménez, A., Cortobius, M., & Kjellen, M. (2014b). Water, sanitation and hygiene and indigenous peoples: A review of the literature. *Water International*, *39*(3), 277–293. doi:10.1080/02508060.2014.903453

Jiménez, A., Jawara, D., LeDeunff, H., Naylor, N., & Scharp, C. (2017). Sustainability in practice: Experiences from rural water and sanitation services in West Africa. *Sustainability*, *9*(3), 403–417. doi:10.3390/su9030403

Jiménez, A., Mtango, F., & Cairncross, S. (2014a). What role for local government in sanitation promotion? Lessons from Tanzania. *Water Policy*, *16*(6), 1104–1120. doi:10.2166/wp.2014.203

Jiménez, A., & Pérez-Foguet, A. (2011a). Water point mapping for the analysis of rural water supply plans: A case study from Tanzania. *Journal of Water Resources Planning and Management ASCE*, *137*(5), 439–447. doi:10.1061/(ASCE)WR.1943-5452.0000135

Jiménez, A., & Pérez-Foguet, A. (2011b). The relationship between technology and functionality of rural water points: Evidence from Tanzania. *Water Science and Technology*, *63*(5), 949–956. doi:10.2166/wst.2011.274

Jiménez, A., & Pérez-Foguet, A. (2011c). Implementing pro-poor policies in a decentralized context: The case of the rural water supply and sanitation program in Tanzania. *Sustainability Science*, *6*(1), 37–49. doi:10.1007/s11625-010-0121-1

Jiménez, A., & Pérez-Foguet, A. (2012). Quality and year-round availability of water delivered by improved water points in rural Tanzania: Effects on coverage. *Water Policy*, *14*(3), 509–523. doi:10.2166/wp.2011.026

Kunneke, R., Groenewegen, J., & Ménard, C. (2010). Aligning modes of organization with technology: Critical transactions in the reform of infrastructures. *Journal of Economic Behavior and Organization*, *75*(3), 494–505. doi:10.1016/j.jebo.2010.05.009

Laffont, J. J. (2005). *Regulation and development*. Cambridge: Cambridge University Press.

Laffont, J. J., & Tirole, J. (1993). *A theory of incentives in procurement and regulation*. Cambridge: MIT Press.

Levy, B., & Spiller, P. (1996). *Regulations, institutions and commitments. A comparative analysis of telecommunications regulation*. Cambridge: Cambridge University Press.

Libecap, G. (2014). Addressing global externalities: Transaction costs considerations. *Journal of Economic Literature, 52*(2), 424–479. doi:10.1257/jel.52.2.424

Lobina, E. (2013). remediable institutional alignment and water service reform: Beyond rational choice. *International Journal of Water Governance, 1*(1–2), 109–132. doi:10.7564/12-IJWG3

Mariz, C., deMénard, C., & Abeillé, B. (2014). *Public procurement reforms in Africa. Challenges in institutions and governance.* Oxford: Oxford University Press.

Marques, R. (2010). *Regulation of water and wastewater services – An international comparison.* London: IWA.

Marques, R., Simões, P., Pires, J. S., & Almeida, J. (2009, September). Regulation of water and wastewater services. An international comparison. In *Water Utility Management International* (pp. 26–29).

Marshall, W., & Weingast, B. (1988). The industrial organization of congress. *Journal of Political Economy, 96*(February), 132–163.

Ménard, C. (2016, March). *Institutional aspects of governance in fisheries management: Five case studies* (Report to OECD (TAD/FISH)).

Ménard, C. (2017). Meso-institutions: The variety of regulatory arrangements in the water sector. *Utilities Policy, 49,* 6–19. doi:10.1016/j.jup.2017.05.001.

Ménard, C. (2012). Risk in urban water reform: A challenge to public–private partnership. In A. Gunawansa & L. Bhular (Eds.), *Water governance: An evaluation of alternative architectures* (pp. 290–320). Cheltenham: Edward Elgar.

Ménard, C. (2008). A new institutional approach to organization. In C. Ménard & M. Shirley (Eds.), *Handbook of new institutional economics* (pp. 281–318). Boston: Springer.

Ménard, C., & Ghertman, M. (Eds.). (2009). *Regulation, deregulation and re-regulation.* Cheltenham: Edward Elgar.

Milgrom, P., & Roberts, J. (1992). *The economics of organization and management.* Englewood Cliffs: Prentice Hall.

North, D. C. (1990a). *Institutions, institutional change, and economic performance.* Cambridge: Cambridge University Press.

North, D. C. (1990b). A transaction cost theory of politics. *Journal of Theoretical Politics, 2*(4), 355–367. doi:10.1177/0951692890002004001

OECD. (2009). *Private sector participation in water infrastructure: OECD checklist for public action.* OECD Studies on Water. Paris: OECD.

OECD. (2011). *Water governance in OECD countries: A multi-level approach.* OECD studies on water. Paris: OECD. Retrieved from http://www.oecd-ilibrary.org/environment/water-governance-in-oecd-countries_9789264119284-en

OECD. (2014). *Water governance in the Netherlands.* Paris: OECD.

OECD. (2015a). *Principles on water governance.* (Coordinated by A. Akhmouch). Paris: OECD.

OECD. (2015b). *The governance of water regulators.* (Coordinated by C. Kauffmann). Paris: OECD. doi:10.1787/9789264231092-en

OECD. (2015c). *Inventory: Water governance indicators and measurement frameworks.* (Collected by A. Akhmouch and O. Romano). Paris: OECD.

OECD. (2015d). *Water resources governance in Brazil.* (Coordinated by A. Akhmouch, X. Leflaive and O. Romano). Paris: OECD.

OECD. (2016). *Water governance in cities.* Paris: OECD. doi:10.1787/9789264251090-en

OECD. (2017). *The political economy of biodiversity policy reform* (Coordinated by K. Dominique and K. Karousakis). Paris: OECD. doi:10.1787/978926429545-en

Ostrom, E. (2005). *Understanding institutional diversity.* Princeton: Princeton University Press.

Ostrom, E. (2014). Do institutions for collective action evolve? *Journal of Bioeconomics, 16*(1), 3–30. doi:10.1007/s10818-013-9154-8

Overseas Development Institute (2015, October). *Private sector and water supply, sanitation, and Hygiene* (Report by N. Mason, M. Matoso and W. Smith). Retrieved March, 2017, from https://www.odi.org/?gclid=CIz-2qOriNMCFdUV0wodr4oNtA

Phumpiu, P., & Gustafsson, J. E. (2009). When are partnerships a viable tool for development? Institutions and partnerships for water and sanitation service in Latin America. *Water Resources Management, 23*(1), 19–38. doi:10.1007/s11269-008-9262-8

Ribot, J. C., Agrawal, A., & Larson, A. M. (2006). Recentralizing while decentralizing: How national governments reappropriate forest resources. *World Development, 34*(11), 1864–1886. doi:10.1016/j.worlddev.2005.11.020

Robinson, J. A. (2010). The political economy of institutions and resources. In D. R. Leal (Ed.), *The political economy of natural resource use. Lessons for fisheries reform* (pp. 45–56). Washington, DC: IBRD/World Bank.

Rusca, M., Schwartz, K., Hadzovic, L., & Ahlers, R. (2015). Adapting generic models through bricolage: Elite capture of water users associations in Peri-Urban Lilongwe. *European Journal of Development Research, 27*(5), 777–792. doi:10.1057/ejdr.2014.58

Savedoff, W., & Spiller, P. (1999). *Spilled water. Institutional commitment in the provision of water services.* Washington, DC: Inter-American Development Bank.

Schlüter, M., & Herrfahrdt-Pahle, E. (2011). Exploring resilience and transformability of a river basin in the face of socioeconomic and ecological crisis: An example from the Amudarya River Basin, Central Asia. *Ecology and Society, 16*(1), 32. doi:10.5751/ES-03910-160132

Sehring, J. (2009). Path dependencies and institutional bricolage in post-Soviet water governance. *Water Alternatives, 2*(1), 61–81.

Shirley, M. (Ed.). (2002). *Thirsting for efficiency.* London: Elsevier-Pergamon/World Bank Group.

Shirley, M. (2008). *Institutions and development.* Cheltenham: Edward Elgar.

Spiller, P. (2009). An institutional theory of public contracts: Regulatory implications. In C. Ménard & M. Ghertman (Eds.), *Regulation, deregulation, reregulation. Institutional perspectives* (pp. 45–66). Cheltenham: Edward Elgar.

Stockholm Environment Institute. (2013) *Sanitation policy and practice in Rwanda: Tackling the disconnect.* Policy Brief. Stockholm. Retrieved from http://seiinternational.org/mediamanager/documents/Publications/SEI-PolicyBrief-SanitationPolicyAndPracticeInRwandaTacklingTheDisconnect-2013.pdf

Tomson, W. (2009). *The political economy of reform. Lessons from pensions, product markets and labour markets in ten OECD countries.* Paris: OECD.

Transparency International. (2008). *Global corruption report 2008: Corruption in the water sector.* Cambridge: Cambridge University Press.

Tremolet, S., & Binder, K. (2010). *The regulation of water and sanitation services in DCs.* Paris: Agence Française de Développement.

Tropp, H. (2007). Water governance: Trends and needs for new capacity development. *Water Policy, 9*(S2), 19–30. doi:10.2166/wp.2007.137

UNDP. (2013). *Users' guide to assessing water governance.* Oslo: UNDP.

UNDP Water Governance Facility/SIWI. (2015). *Water governance in perspective.* Retrieved from http://watergovernance.org/resources/water-governance-in-perspective/

UNDP Water Governance Facility/UNICEF. (2015). *Accountability in WASH: Concept note* Accountability for Sustainability Partnership: UNDP Water Governance Facility at SIWI and UNICEF. Stockholm. Retrieved from http://www.watergovernance.org/Accountability-for-Sustainability

UNESCO. (2006). *The United Nations World Water Development Report 2: Water, a shared responsibility.* Paris: UNESCO. Retrieved from http://www.unesco.org/bpi/wwap/s/

UNICEF/WHO Joint Monitoring Program. (2015). *2015 update and MDG assessment.* Retrieved from https://www.wssinfo.org/fileadmin/user_upload/.../JMP-Update-report-2015_English.pdf

Vatn, A. (2015). *Environmental governance. Institutions, policies and actions.* Cheltenham: Edward Elgar.

Warner, J., Butterworth, J., Wegerich, K., Mora Vallejo, A., Martinez, G., Gouet, C., & Visscher, J. T. (2009). *Corruption risks in water licensing with case studies from Chile and Kazakhstan* (Swedish Water House (SIWI) Report 27).

Water Integrity Network. (2015). *Water integrity global outlook.* Berlin. Retrieved from www.waterintegritynetwork.net/wigo/

Williamson, O. E. (1985). *The economic institutions of capitalism*. New York: Free Press/ Macmillan.
Williamson, O. E. (1996). *The mechanisms of governance*. Oxford: Oxford University Press.
World Development Report. (2017). *Governance and the law*. Washington, DC: World Bank. Retrieved from http://www.worldbank.org/en/publication/wdr2017
World Water Development Report. (2006). *Water a shared responsibility*. Paris: UNESCO. Retrieved from http://www.unesco.org/new/en/natural-sciences/environment/water/wwap/wwdr/wwdr2-2006/

OPEN ACCESS

Stakeholder engagement in water governance as social learning: lessons from practice

Uta Wehn, Kevin Collins, Kim Anema, Laura Basco-Carrera and Alix Lerebours

ABSTRACT

The OECD Principles on Water Governance set out various requirements for stakeholder engagement. Coupled with conceptualizations of social learning, this article asks how we define and enact stakeholder engagement and explores the actual practice of engagement of stakeholders in three fields of water governance. The results suggest that a key consideration is the purpose of the stakeholder engagement, requiring consideration of its ethics, process, roles and expected outcomes. While facilitators cannot be held accountable if stakeholder engagement 'fails' in terms of social learning, they are responsible for ensuring that the enabling conditions for social learning are met.

Introduction

A core principle of the Water Governance Principles formulated by the Organisation for Economic Co-operation and Development (OECD) in 2015 recognizes the importance of promoting stakeholder engagement in water governance processes (OECD, 2015). Stakeholder engagement has become a central requirement for water-related projects in many different contexts, amid demands for long-term benefits such as sustainability and resilience as well as developing fragile but powerful intangible assets such as trust, ownership and acceptability (Von Korff, Daniell, Moellenkamp, Bots, & Bijlsma, 2012).

While international and regional treaties such as the Aarhus Convention or the Dublin Principles for Integrated Water Resources Management require citizen participation and the establishment of mechanisms for public participation in decision making, the importance given to these institutional imperatives, their interpretation and the extent of their implementation varies. Several studies have also shown that many participatory approaches fail to lead to more informed and effective policy and practice (Behagel & Turnhout, 2011; Edelenbos & Klijn, 2006; GWP, 2000), whether from insufficient or misused resources, organizational intransigency or poor design for those processes to deliver their full potential. Also, groups such as young people and local community members are often overlooked due to power differentials and organizational 'expectations'. Thus participation is often poorly defined, too often considered a 'formality' or an adjustment variable within the budget of scoping studies (like environmental and social impact assessments), such that an obligation to participate is rarely taken as an opportunity to improve projects and generate collective learning (Barreteau, Bots, & Daniell, 2010; Irvin & Stansbury, 2004; Rinaudo & Garin, 2005).

Despite these concerns, stakeholder engagement, as a subset of broader participatory imperatives, is often a requirement for policy makers, authorities or utilities to engage with citizens or members of the public who have a stake in the decision or outcome, but may not normally be considered part of the core decision-making process. However, this interpretation can overlook the equally important role of decision makers and practitioners as stakeholders and the possibility of their engagement with each other, leading to learning and improvements and increased likelihood of situation improvement.

Despite the claimed value that sound stakeholder engagement can provide to water projects (OECD, 2015), the business case remains hard to defend for promoters of dialogue, when costs are immediately measurable but benefits could take time to arise, remain opaque and be unequally distributed among stakeholders. In a recent OECD survey of 215 water stakeholders, only 8% perceived market opportunities as a driver for stakeholder engagement, compared to regulatory or emergency-related drivers, suggesting that stakeholder engagement, while important as a regulatory requirement, is a low business priority in terms of economic development and business growth. In warning of a 'ticking the box' approach to stakeholder engagement, the survey reveals that stakeholders mostly interact within their immediate sphere of activity, that engagement processes are rarely evaluated, and that there is no simple way to measure their impact (OECD, 2015). Thus, stakeholder engagement varies in conceptualization, drivers, 'fit' with organizational cultures and goals, and practice, with commensurate variation in outcomes and interpretations of viability and usefulness. How then should we conceptualize and enact stakeholder engagement?

To improve understanding of Principle 10 of the Principles on Water Governance, 'Promote stakeholder engagement for informed and outcome-oriented contributions to water policy design and implementation' (OECD, 2015), this article aims to contribute to theoretical and empirical debates on what stakeholder engagement means and may deliver in practice in the context of water governance. Specifically, we argue that stakeholder engagement entails not only public participation but multi-stakeholder interaction, dialogue and learning and that it requires more than a top-down decision-making process to make it succeed. Drawing on conceptualizations of social learning as well as participation in decision making, this article reflects on different modalities for the engagement of distinct stakeholders in three fields of water governance: the use of citizen observatories of water to engage citizens in flood-risk management in the UK, the Netherlands and Italy; the involvement of policy makers and practitioner members in the catchment-based approach as part of implementing the Water Framework Directive in England; and the role of under-represented groups, i.e. young people, the homeless and local communities, in water security at different levels —regional (in Europe), district (in France), and community (in Kenya). The empirical evidence is subjected to within- and across-case analyses to generate in-depth understanding of the dynamics of different stakeholder engagement processes in these respective fields of water governance.

The article is structured as follows. The second section defines the theoretical context of our research, followed by the presentation of methodological details in the third section. The results of the empirical research per case study are analyzed in the fourth section. Using the conceptual framing of stakeholder engagement as social learning, the fifth section discusses the findings and lessons from practice, and the sixth proposes concrete recommendations. The seventh section offers conclusions.

Theoretical context

Stakeholder engagement in decision making

Participation as a concept, method and practice has been discussed extensively in the literature since Arnstein's (1969) ladder offered a simple structure for identifying power-based degrees of citizen involvement in decision making (Bruns, 2003; Collins & Ison, 2009; Fung, 2006; Hurlbert & Gupta, 2015; Ison, Röling, & Watson, 2007; Mostert et al., 2007; Reed, 2008; Voinov et al., 2016). We do not rehearse these debates here, other than to note that they highlight the importance of clarity on meanings and concepts. There are distinctly different forms of participation, with varying outcomes and impacts (Fung, 2006; Reed, 2008) that depend on the contextual setting and the nature of the issue or problem at hand (Hurlbert & Gupta, 2015). This article follows Rowe and Frewer (2004, p. 253) in recognizing participation, in broad terms, as 'the practice of involving members of the public in the agenda setting, decision-making, and policy-forming activities of organizations/institutions responsible for policy development'. They go on to distinguish public *consultation* as characterized by an active process of information exchange and dialogue between those involved. They suggest public *engagement* as a collective term encompassing public communication, public involvement and public participation (p. 254) – the distinctions of each being dependent on the dynamics of information flow between the sponsor of the process and the participants and on the effectiveness of the mechanisms deployed commensurate with these distinctions. Thus, engagement is a wide-ranging, but active, dynamic process where stakeholders are 'allowed in' to participate in decision-making processes.

The OECD (2015, p. 32) defines engagement as a broad umbrella term and stakeholder engagement as the opportunity for those with an interest, or 'stake', to take part in decision-making and implementation processes. Here, stakeholders are distinct from simply the wider 'public' and can also include government actors, the private sector, regulators and NGOs.

A stakeholder is usually defined as someone having an interest in a particular situation, even if this interest is not recognized or acknowledged by others. Nevertheless, awareness of the dynamics of engagement leads some authors (Collins, Blackmore, Morris, & Watson, 2007; SLIM, 2004a) to suggest that *stakeholding* may be a preferable concept because it conveys the notion that stakeholders *actively* construct, promote and defend their stake over time and can sometimes defend their stake and exert influence by *not* engaging in participatory processes. A focus on *stakeholding* as a process rather than *stakeholder* as a noun allows insights into how stakes are constructed, reshaped and 'shared' in social learning processes (see below).

Stakeholder engagement is seen as a means of contributing to improved water governance, where governance is defined as the policy and practices giving rise to particular forms of water managing in different contexts. It is defined as a critical principle for sustainable development and building a resilient society (Gunderson, 2003) and is both a means and an end, insofar as it can lead to increased stakeholder empowerment and make planning and decision-making processes more transparent and democratic (Hare, Letcher, & Jakeman, 2003). It is also claimed to enhance the capacity of individuals to improve their own lives, facilitating social change (Cleaver, 1999). Local knowledge and expertise can be valuable for understanding local situations and contexts, planning objectives and policy measures, as well as improving and/or creating innovative and alternative strategies; as a result, the sustainability of the adopted policy strategy will generally be higher (Hurlbert & Gupta, 2015). Stakeholder engagement can also promote social learning, as stakeholders *acquire* (rather than just convey) knowledge and collective skills through better understanding of their situation as well as the perceptions, concerns and interests of other stakeholders (Basco-Carrera, Warren, Van Beek, Jonoski, & Giardino, 2017; Collins & Ison, 2009; Evers et al., 2012; Hare, 2011; Voinov & Bousquet, 2010). Finally, stakeholder engagement can foster consensus among competing organizations by opening channels of communication, generating mutual understanding, and negotiating alternative solutions (Loucks, Van Beek, Stedinger, Dijkman, & Villars, 2005; Sadoff & Grey, 2005; Hare, 2011; Zeitoun & Mirumachi, 2008).

In this vein, the OECD principle of stakeholder engagement is aimed at enabling informed and outcome-oriented contributions to water policy design and implementation (Akhmouch and Clavreul, 2016). The OECD sets out various requirements for stakeholder engagement, which in summary are: recognizing the range of actors with a stake in a situation and understanding their possibly diverse responsibilities; paying special attention to underrepresented groups; identifying the process of decision making and stakeholder inputs; encouraging capacity development of stakeholders; assessing and evaluating engagement processes; promoting conducive institutions; and contextualizing stakeholder engagement initiatives.

However, the actual dynamics in the engagement process mean that these positive outcomes are far from automatic or guaranteed, as increasing evidence shows (Behagel & Turnhout, 2011; Edelenbos & Klijn, 2006; Furber, Medema, Adamowski, Clamen, & Vijay, 2016). Stakeholder involvement implies – explicitly or implicitly – trade-offs in terms of representativeness, inclusion, or (in)equality in interactive processes (e.g., Sørenson, 2002; Mayer, van Bueren, & Bots, 2005; Sørenson & Torfing, 2007), i.e. between the 'breadth' and 'depth' of involvement (Voinov et al., 2016). For example, in terms of the breadth of stakeholders involved or procedural fairness (Adger, Paavola, Huq, & Mace, 2006), under-represented groups such as young people, local communities and the homeless are not frequently acknowledged as 'well-placed' stakeholder groups, e.g. to address the challenges related to water security. Their relatively limited experience, knowledge or vulnerability make it difficult for them to be considered as key stakeholder groups, and therefore to participate actively in decision-making processes. Substantive aspects also come into play (Van Buuren, Driessen, Teisman, & Van

Rijswick, 2014) concerning the extent to which all stakeholder inputs and interest have actually been taken into account. In water management and spatial development, policy making and decision making have tended to be expert-driven and expert-produced according to technocratic standards (DeSario & Langton, 1987; Fischer, 2000; Hisschemöller, 1993). This includes the belief that the desirability of the solution can be shown by standardized methods and technical procedures and that the use of available expert knowledge is sufficient for an efficient implementation of the solution. Consequently, the participation of stakeholders is often considered superfluous, because they do not have the (technical) knowledge and expertise required for situation appraisal or resolution (Edelenbos, Van Schie, & Gerrits, 2008). Moreover, the dynamic of stakeholder engagement is changing and increasingly subject to intermediation via digital innovations (Voinov et al., 2016; Wehn & Evers, 2015). Evidence of how to capture these emerging opportunities as well as how to address the accompanying challenges is limited, not least due to a lag in updating stakeholder conceptualizations for the digital age (Wehn & Evers, 2015). Nevertheless, while important, the mechanism of stakeholder engagement is secondary to the underlying purpose. In this sense, emphasis is moving away from procedural nicety to a fundamental concern with making sense of complex situations, where stakeholder engagement is seen as a form of social learning.

Social learning

Social learning has become an increasingly frequent term in the literature on participation and stakeholder engagement processes, but its interpretation, use and endorsement vary (Blackmore, Ison, & Jiggins, 2007). Initially coined by Bandura (1977) to describe *individual* learning in a social context, the concept of social learning has since been expanded to include learning emerging from *collectives* or groups (Ison et al., 2007; SLIM, 2004b). There are many authors exploring the concept of social learning in environmental policy and water governance contexts (Colvin et al., 2014; Ison et al., 2007; Pahl-Wostl & Hare, 2004; Pahl-Wostl et al., 2008; Röling, 2002; Scholz, Dewulf, & Pahl-Wostl, 2014; Woodhill & Röling, 1998). Some of these authors are exploring environmental problems based on a more integrative approach and systemic understanding using systems approaches. These aim to engage with the inherent complexity of water governance and how change in the 'right' direction can be fostered via social learning, i.e. fostering the capacity to becoming adaptive systems (see e.g. Ison, Collins, & Wallis, 2015). We do not rehearse these complex debates and differences here, but note that the common element in this discourse is the realization that complex environmental situations require, among other factors, collective learning and common understanding.

One of the most salient aspects of social learning related to stakeholder engagement is therefore the collective – rather than individual – process of learning, knowledge co-creation and accumulation of wide experiences to generate a broader knowledge and evidence base from which decisions can be taken. Specifically, we consider social learning an emerging governance mechanism to promote concerted action among stakeholders to improve water governance (Collins et al., 2007; Ison et al., 2007). Concerted action is framed not as a replacement, but as complementary or enhancing to existing mechanisms, e.g. regulations, fiscal measures and education, through for example information provision (SLIM, 2004b).

In this sense, social learning can be understood and summarized as one or more of the following (after Collins & Ison, 2009):

- The convergence of goals (expressed as purpose)
- The process of co-creation of knowledge which provides insights into the causes of a situation and the means of its possible transformation
- The changes in behaviours and actions resulting from new understandings
- An emergent property of the process to transform a situation.

The implications of this interpretation of social learning are that stakeholder engagement is conceived as purposeful and designed to enhance cooperation and learning between stakeholders. This enables understanding of the water governance situation and how it can be progressed and transformed, including changes in mental models, beliefs, perceptions, and – as a result – practices. Increasingly, attention also turns to processes of social learning mediated by online environments (Joshi and Wehn, 2017; Voinov et al., 2016; Wehn & Evers, 2015), which are subject to change and evolution triggered by differing ways in which data and information can be shared and knowledge co-created. Online portals (Bourget, 2011; Cockerill, Tidwell, Passell, & Malczynsky, 2007; Evers et al., 2012; Jonoski & Evers, 2013) and social media (Wendling, Radisch, & Jacobzone, 2013) are gaining relevance as alternative and mainstream mechanisms for e-participation.

With new forms and increased opportunities for engaging stakeholders, we investigate stakeholder engagement as social learning in specific cases in order to inform and guide the implementation of the OECD principle of stakeholder engagement. Specifically, we aim to guide the assessment of the process and outcomes of stakeholder engagement in terms of social learning, focusing on stakeholder dynamics, knowledge co-creation, and individual behavioural changes as well as collective transformation.

Methodology

Research design and selected case studies

The research reported here was not designed or undertaken as part of a single project or initiative, and not intended specifically to illuminate the OECD principles. Instead, the authors have collaborated post-research to explore insights and findings which may bear on understanding and furthering stakeholder engagement as social learning in water governance. The designs of the case studies, their focus and their contexts thus vary considerably, as shown in Table 1.

While the diversity of design is evident, the focus on social learning is a common theme, albeit not necessarily understood in advance in each project with reference to the framing of social learning noted above. The analysis therefore proceeds first with reporting the main findings in the context of each case study and then, second, on an ex post basis to develop a retrospective meta-analysis of the findings emerging across the case studies and the extent to which these findings have bearing on the OECD principle.

Table 1. Overview of empirical research for the case studies.

	Catchment-based approach for the Water Framework Directive	Water Youth Network water projects	WeSenselt citizen observatories
Geographic focus of the case	UK	Europe (consultation during Stockholm World Water Week), France (Paris), Kenya (Tigithi community, Laikipia District)	Doncaster, UK; Delfland, Netherlands; Alto Adriatico, Italy
Scope	National	Regional, district, community	City, region, catchment
Data collection instruments	Workshops and interviews	Focus group discussions, questionnaires and interviews	Interviews and focus group discussions
Timing of empirical research	2013–16	2015–2016	2012–2016
Type of respondents/ interviewees	Policy makers from government and NGO communities; practitioners; researchers	Under-represented and vulnerable groups (young people, homeless and community members)	Authority representatives (policy and decision makers), trained volunteers, general public in Doncaster, Vicenza and Delfland
No. of respondents/ interviewees	Two workshops of about 15 participants each, plus five interviewees in 2015	245 respondents over the course of 2 years: 91 persons consulted in small focus discussions, 20 interviewees, and 134 questionnaire respondents	83 interviewees over the course of three years

Data collection

As would be expected in diverse projects with diverse aims, scales and stakeholders, the processes of data collection varied across the case studies. The generated evidence is anecdotal, but within an action research context and process, and provides a rich basis for our inquiry into the process and outcomes of stakeholder engagement in terms of social learning. A summary of the data collection processes is shown in Table 1. In our view, rather than being a weakness, the diversity offers the opportunity to explore the role of social learning in stakeholder engagement in a range of contexts and situations and thus whether the OECD principle has relevance.

The main element of the UK case study reported here focusses on water governance in relation to implementation of the Water Framework Directive in England. In the light of ongoing concerns that current institutional arrangements are insufficient to deliver improvements in water quality (Watson, 2014), in 2012 the Department for Environment, Food and Rural Affairs (DEFRA) initiated a Catchment-Based Approach (CaBA) to fill a policy and practice 'gap' between the regional river basin (consisting of several catchments) and the individual water body focus of the Water Framework Directive – an arrangement which hitherto had largely ignored individual catchments. With policy leadership and seed funding from DEFRA and support from the Environment Agency of England and Wales, CaBA has developed into a network of over 100 catchments in England and Wales adopting a community-based approach to improve water environments. Over the last four years, a team of researchers at the Open University in the UK has been undertaking action research work with the national CaBA National Steering Group (NSG). With the CaBA initiative now in place, what does more integrated and systemic water governance look like? It was with this question that the Open University researchers, as part of the CADWAGO project, engaged the policy makers and practitioner members of the CaBA steering group and associated stakeholders. The research reported here centres on a systemic co-inquiry with stakeholders to improve understanding and practices in relation to water governance.

Informed by traditions in systems theory, methods and approaches, systemic co-inquiry is a mode of investigation that is open and flexible as to the nature of the situation of concern, the direction of the inquiry and the different epistemologies and methodological traditions of the stakeholders involved (Ison, 2010). At the core of the co-inquiry is a commitment by the participants (including the researchers) to a social learning process, in this case using elements of soft systems methodology (SSM) (Checkland, 1981; Checkland & Scholes, 2002) and diagramming skills taught and developed at the Open University. The outcomes of a systemic co-inquiry are not predetermined but centre on stakeholders' learning and possible changes in understanding about a situation which can lead to new forms of policy and practice.

The WeSenseIt case studies were undertaken within an action research framework (Greenwood & Levin, 1998; Lewin, 1946), triggering change – in this case the participatory development of the citizen observatories over the course of four years – while at the same time studying and reflecting on the wider effects and outcomes that are being generated. The researchers thus had dual roles as project team members and social scientists studying the emerging changes and capturing changes in behaviour of the stakeholders in the local water governance processes, on the basis of interviews and focus group discussions with citizens and local authorities. These data collection efforts were undertaken according to the conceptual frameworks adopted for WeSenseIt (Wehn & Evers, 2015; Wehn, McCarty, Lanfranchi, & Tapsell, 2015; Wehn, Rusca, Evers, & Lanfranchi, 2015).

The Water Youth Network (WYN) projects were led in 2015 and 2016, under several project funding and project leaders. Each has its own evaluation and research components. An analysis was then conducted to assess the components of young people and vulnerable people's participation and mobilization. The project leaders and researchers were consulted and their inputs were consolidated by two young researchers of the WYN.

Data analysis

For the UK case, data analysis was largely undertaken by the participants themselves during the co-inquiry events as they worked through a range of tasks to develop their thinking. Post-workshop analysis was mostly in the form of writing up the results in a readable form: a workshop report. The diagrammatic elements of the workshop (e.g. systems diagrams) are not readily analyzable, beyond the role they serve in the discussion and development of ideas.

For the WeSenseIt cases, interviews and focus group discussions were transcribed and initially analyzed according to the conceptual frameworks adopted for WeSenseIt. The collected information was analyzed by attributing the focus group transcripts, texts from the interviews and observations to different predefined indicators of good governance, community resilience and participation in general. This generated a structured matrix with qualifications and quotes per indicator per case.

For the WYN cases, data analysis was done at the end of each project, based on data collected before, during and after the activities. Data were collected through key stakeholders' interviews, consultation and household surveys. A secondary data analysis was conducted in 2016. All documents from the projects, feedbacks from team leaders and project researchers were used.

The meta-analysis undertaken for this article took the form of the contributing authors identifying learnings from their respective case and then identifying emerging themes in relation to the conceptualizations of social learning used in this article. An initial tabulation served as a device for refining our collective understanding of, and discussion about, the commonalities and differences between the cases and how the lessons learned can inform the implementation of the OECD principle of stakeholder engagement.

Results and analysis

This section presents a detailed analysis of the selected cases, covering the following elements for each case to capture the dynamics of stakeholder engagement and social learning: (1) the 'who, what, where, when, how and why' of stakeholder engagement; (2) the extent of the convergence of goals during the social learning process; (3) the process of co-creation of knowledge which provides insights into the causes of a situation and the means of its possible transformation; (4) the changes in behaviours and actions resulting from new understandings; and (5) the extent to which social learning is an emergent property of the process to transform the situation. In terms of correlating these to the OECD principle of stakeholder engagement, we suggest that these elements are key to understanding the *purpose* of stakeholder engagement and thus its role as a means to improve outcome-oriented contributions to water policy design and implementation. In each of the cases noted below, we have described the diverse actors involved and explored their motivations and interactions. Capacity development has been a key design element of each and was undertaken in response to a variety of contexts. As these are mostly research-driven cases, our assessment of the cases and the contributions of stakeholder engagement is focussed on the emergent learning rather than detailed cost basis.

Catchment-based approach (UK)

Who, what, where, when, how and why?

This case focusses on the social learning arising from two events undertaken as a part of the systemic co-inquiry and reported more fully in Foster (2017). The two events in 2016, each with 15 representatives from senior policy and practitioner communities, focused on (1) the current and (2) the future water governance situation in England, respectively, and each provided a mix of participatory sessions and presentations exploring aspects of water governance and possible improvements and opportunities for concerted action.

Convergence of goals

The act of initiating the co-inquiry set in train at least a collective interest in the situation: how CaBA can contribute to managing water more effectively in England. Negotiations and exploration of possible themes with DEFRA and the chair of the CaBA NSG helped give some shape to the inquiry, but the exact focus was left open. In this sense, the inquiry did not require convergence of goals from the outset or as a precondition, other than a willingness of participants to engage in the process itself.

Process of knowledge co-creation

The first participatory activity in the first event – focussing on the current water governance situation – used a technique of rich pictures. This is a visual/diagrammatic exploration of the situation in unstructured form and is part of SSM. The resulting rich pictures (see Foster, Collins, Ison, & Blackmore, 2016) reveal agreement on the complexity and messiness of the situation, showing, for example, conflicting stakeholder interests; crises, including flooding and pollution; and the diverse and sometimes overlapping roles of institutions and organizations at different levels, from local to EU. In comparing their rich pictures, stakeholders realized they had difficulty in gaining an overall understanding of the water governance system – not least because of the problem of determining the nature of the system or its constituent boundaries.

There were some significant areas of consensus, leading to an agreed description of the current water governance system and its stakeholders in the form of a 'root definition' (a methodological convention in SSM). Paraphrased, the definition of the current water governance system described it as a disconnected and opaque system, managed by the EU, government and water companies, to deliver public water supply using top-down regulatory approaches in order to support economic growth and welfare (Foster et al., 2016).

In this case, the root definition revealed collective insights into the form and purpose of the current water governance system and its key limitations and thus the basis for which improvements could be discussed and identified.

Suggestions for possible improvements to existing water governance took the form of 'what is'/'what ought to be' statements – each identifying a key aspect of the system and ways it could be improved. Through negotiation and discussion, the stakeholders identified a shared concern that the Water Framework Directive's ecological objectives were being pursued at the expense of a wider range of social, environmental and economic concerns which might also lead to improvements in water quality, but, critically, as an *emergent property* of an improved water governance system rather than a predetermined target.

By the end of the first event, the stakeholders had identified a range of shared concerns about the purpose and boundary of current water governance in England, including the role of CaBA, and had also begun to explore the threads of possible improvements.

The second event then focussed on future water governance, again using SSM to structure the inquiry, beginning with a rich picture. Although retaining the complexity and messiness, the rich pictures of future water governance depicted water governance as a more positive imperative, with different stakeholders working towards shared goals and an emphasis on social/community-led learning and action, shared ownership and responsibility, and collaboration. This shift was also very evident in the root definition of a future water governance system. This foresaw an iterative, place-based, more reflexive, learning system to optimize the management of water by engaging and empowering society to make equitable decisions and take collective/concerted actions; one which valued natural capital in order to deliver human health and well-being within the constraints of social, environmental and economic capital (Foster et al., 2016). The discussions are notable because of *the lack* of specific focus on water quality as the prime target or goal.

In SSM, root definitions convey a sense of an ideal state to offer comparison with the 'real world' in order to identify scope for improvements and actions. In this regard, determining root definitions becomes a social learning process among the stakeholders, and the root definition is both a device for, and evidence of, social learning about possible improvements to effect a transformation. Even so, it is important to note that not everyone was always entirely comfortable with the process or the outcome, particularly the limited time available for discussion and detailed analysis.

Changes in behaviour and actions resulting from new understandings

If we recognize a dynamic relationship between changes in thinking and changes in behaviour, it is important to note that although these were part of an ongoing relationship with the various stakeholders involved, the co-inquiry events described here were time-constrained and one-off processes. As such it is not possible to determine long-term changes in behaviour without a longitudinal data-set. However, for the participants, the workshop processes represented a marked change in how ideas about water governance *could* be approached, discussed, explored and different perspectives acknowledged. Remarks to the research team during and after the co-inquiries suggest that the discussions were of significance in shaping their ideas and follow-up discussions in thinking about CaBA as an innovation in water governance and its direction of travel.

An emergent property of the process to transform the situation

An emergent property arises from the interaction of various parts of a system. In this case, knowledge and insights into the boundaries and constraints of current water governance systems and the different framings of water governance might be considered an emergent property made possible by the social learning process described above. By engaging with each others' multiple perspectives, stakeholders were able to develop new perspectives and, as a result, appreciate and identify new framings and forms of water governance. Nevertheless, the extent to which this emergent property 'holds' or has longevity remains in question.

WeSenseIt citizen observatories of water

Who, what, where, when, how and why?

Between 2012 and 2016, an EU-funded FP7 project called WeSenseIt designed and implemented three citizen observatories to test, experiment and demonstrate their potential: involving citizens and not just scientists and professionals in data collection and establishing a two-way communication paradigm between citizens and authorities involved in flood-risk management. Each was fed with data by both physical sensors (e.g. water-level sensors) and social sensors (e.g. mobile applications). All three observatories focused on flood risk and were put in place in collaboration with water management and/or civil protection agencies. Citizen observatories present the potential for not only higher information density for environmental management but also for considerable improvements in, or in fact new means of, engagement. Their features can enable a two-way communication paradigm between citizens and decision makers, potentially resulting in profound changes to existing flood-risk management processes (Wehn, Rusca, et al., 2015). In collaboration with the respective water management and/or civil protection agencies, Doncaster (UK), Vicenza (Italy) and Delfland (Netherlands) were selected to implement citizen observatories focussed on flood risk. Apart from the local authorities, the observatories aimed to involve citizens, and two of the three observatories involved trained volunteers as well.

The research objective was the same in all three cases: to explore how or under what conditions the combination of new and existing sensing and monitoring technologies, together with interactive information and communication technologies (ICTs) provided in the observatories, can serve to foster e-participation in flood-risk management. Still, over the course of four years, the ways the three observatories were designed, implemented and used evolved in three very different ways. In the Doncaster case, the authorities decided to give the community ownership of the observatory. The online platform got a peer-to-peer focus and was used by the authorities only to monitor the situation on the ground. In the Vicenza case, the authorities took the opposite stand and kept full responsibility for what was posted, when and by whom. This online platform became a tool for coordination and communication between trained volunteers and emergency services. Here the platform was so successful that it was implemented not only in the case study area but at a regional level, with the embedding of the concept of citizen observatories into the regional policy as a means for environmental resources management. Finally, in the Delfland case, the decision was made to build on existing communication structures to ensure responsiveness. The online platform in this case saw little activity.

The convergence of goals

During the set-up of the observatories in both Vicenza and Doncaster, the respective objectives of the authorities and citizens increasingly converged. In Vicenza, existing trained volunteer groups were aiming to enhance their mutual collaboration, where the Civil Protection Agency was mainly interested in more and better 'eyes and ears on the ground'. The observatory platform launched under WeSenseIt redefined working relations (and methods) while also generating more detailed situational awareness.

In Doncaster, the objectives at the start of the project were formulated quite abstractly, and the implementation of the platform was not really successful until the council staff interacted with the engaged flood wardens face to face and demonstrated their understanding of concrete issues on the local and household level. The moment the flood wardens were able to 'put a name to a face' and felt understood, they were also more ready to help out and upload information about their local situation. In both cases, the convergence was triggered by the need or wish of the authorities to get (inter) action on the platform. The citizens and volunteers were accommodated in their objectives and drivers in order to get them to participate.

In Delfland, the citizens targeted for participation were not uniform or organized enough for the water board to converge to their goals, and the observatory in Delfland struggled for participating audiences throughout the project.

The process of knowledge co-creation

One of the first and most valued outcomes in all three cases was the levelled access to relevant and specific information between stakeholders. Authorities can now make use of the eyes and ears of citizens, while citizens and other stakeholders have gained insight into the data that decisions are based on (pumping regimes, road closures, etc.). This emancipates the dialogue and enhances the effectiveness of participation.

One specific example of co-created knowledge comes from the Doncaster case, where at one location the local residents argued that, with previous flood events, one specific field had always been inundated first. The prevailing flood model for that location did not support that observation, but one of the physical sensors related to the platform was placed in the middle of the field. The historic observations of local residents were correct, and the flood model was proved wrong. Using this knowledge hydrologists discovered a slight tidal influence on the behaviour of the river – and with that a few of the candidate flood-prevention measures were proven irrelevant.

An important side effect of this process of co-creation between authorities and citizens was the mutual respect and understanding that slowly grew when they were working together towards the same objectives.

Changes in behaviours and actions resulting from new understandings

The observatories facilitated dialogue between stakeholders, but could not force this dialogue into existence. For all involved parties, feedback on their efforts was needed for them to stay engaged and keep sharing. The success and use of the observatories therefore depended on whether and how the involved authorities wanted to cooperate with citizens and other stakeholders. This is not an easy decision, since this cooperation comes with responsibilities, in terms of both continuity and responsiveness (Wehn, McCarty, et al., 2015).

In all three cases, the authorities appeared hesitant to transfer their interactions with citizens into the online environment of the observatory, owing to fears of interrupting established procedures and the perceived objective of having to respond to citizens' online activities. Liability and accountability concerns are particularly salient in the preparation, impact and response phases of flood-risk management (e.g. having to respond quickly to online posts about flooding, creating an additional channel for the emergency response team, separate from their existing decision support systems). Different choices in that respect were made in the three case studies, leading to the above-mentioned very different outcomes of the observatories.

The closer connection between authority representatives and flood wardens in Doncaster led to less-formal contact; council staff would occasionally attend social events in the neighbourhood, and community members said they were less hesitant to pick up the phone and report or ask something.

In Vicenza, professionals stated that their esteem for the volunteers was already high and had not changed much in that respect. Still, even there, the Civil Protection Agency did keep its hierarchical status but also adopted a more central role between the different volunteer groups.

An emergent property of the process to transform the situation

The results of the observatories' respective flood-risk management strategies focused on sharing information and building community trust; competences in the community and effective response; or efficient and effective risk mitigation. Each of these three outcomes was valued in its own local context by local stakeholders. This shows that citizen observatories are tools or frameworks that can help generate or support an array of participation approaches, depending on how they are put to use.

The three WeSenseIt cases demonstrated how acutely aware authority representatives already are of the responsibilities that go hand in hand with engaging with their citizens. The belief that citizen engagement should be done only if 'you're in it for the long run' is widely shared among the authorities involved in these observatories. The self-imposed standards for responsiveness are very high – almost paralyzing. Trust, ownership, continuity and responsiveness are indeed important issues to take into account. And given different traditions, cultures and backgrounds, these issues would need to be resolved differently in each case.

In each context of the WeSenseIt case studies, tailor-made approaches were chosen, leaving room for citizens and authorities to find their own form of collaboration and mutual trust. In all three cases, the project ended with a mutual understanding of the topic. Futhermore, there is clear evidence that shared understanding and better collaboration can lead to more trust between the parties involved.

The Water Youth Network case

Who, what, where, when, how and why?

In 2015 and 2016, the WYN led a set of consultations and operational and research projects in collaboration with other young people and youth-led organizations to support shaping the 2030 Agenda for Sustainable Development and enhance their participation in the co-creation of knowledge in the water sector regarding securing water for all.

In Sweden, during the Stockholm World Water Week of 2015, a session was organized by two youth organizations: WYN and the World Youth Parliament for Water. As part of the session, a consultation was held to define the role of young people regarding the OECD water governance principles, with special focus on stakeholder engagement (i.e. young people and under-represented groups), and how these principles could be used for youth-led operational water projects. A total of 91 respondents from all over the globe participated in the consultation: 15% were underage people (15–18 years old), 26% were young water students (18–25), 38% were young professionals (25–35) and 21% were senior professionals (over 35).

Multi-stakeholder engagement and partnerships are considered critical elements in the achievement of the Sustainable Development Goals (SDGs). In 2016, the Youth in Action for Sustainable Development Goals (YiA4SDG) project under the ERASMUS+ umbrella aimed to promote the SDGs via a global youth-led call for action and to support practical solutions via awareness-raising, capacity development, co-creation of knowledge, and project piloting and evaluation. A total of 17 pilot initiatives in various countries in Europe, each promoting and helping in the implementation of the SDGs, were supported under the umbrella of the YiA4SDG project.

In line with the OECD water governance principle of stakeholder participation, the Sendai Framework for Disaster Risk Reduction calls for the participation of relevant stakeholders in strengthening disaster-risk governance to manage disaster risk. National and international meteorological agencies predicted a weak El Niño in early 2014. However, by the end of 2014 and early 2015, there was agreement across most agencies that there would be a strong El Niño effect in various regions across the globe, alerting communities and governments to prepare for the worst: excessive rains, drought, flooding and other weather-related disasters. With the research study Local-Level Preparedness and Response to the El Niño Phenomenon Early Warning in Tigithi, Kenya in 2016, the WYN aimed to demonstrate the role that young researchers can have in managing disaster risk by collecting scientifically relevant information to understand local-level preparedness and response to the El Niño–influenced weather in Kenya, as well as empowering local communities as a means towards building a resilient society. A team of seven young researchers (one international and six local) with expertise in disaster-risk reduction analyzed the relevance and usefulness of the forecast advisories and preparedness and response plan for the El Niño phenomenon for the agro-pastoral communities in Laikipia County and their day-to-day challenges. The evaluation methods were surveys and face-to-face interviews with local community members. A total of 134 people were interviewed: 85% adults (over 35 years old; 46 male and 67 female) and 15% young people (7 male and 14 female). Only 40% had basic education; 63% were engaged in farming as their main occupation.

Convergence of goals (expressed as purpose)

The complexity associated with the political and institutional contexts due to differences in ambitions, values and interests between agencies and institutions, different understandings of the problems at stake, or restricted resources often restricts the involvement of under-represented groups. The participation of young people and local communities in decision-making processes and global agendas is therefore frequently considered an added risk rather than an added value. Highlighting their vulnerability and limited experience or knowledge, among other factors, is a commonly used mechanism for restricting their engagement. The WYN aims to shift this paradigm by facilitating the participation of under-represented groups, mainly young people, via a structured engagement process that can accommodate the complexity of the particular context and provides room for the co-creation of knowledge. This framework based on social learning helps reduce risks by allowing under-represented groups to better understand the complexity and functioning of a particular governance and decision-making context. It also facilitates the constructive contribution of these groups by means of local knowledge, new developments and innovative ideas.

The process of knowledge co-creation

The three WYN projects presented here enabled social learning for the participants, by providing them with knowledge, access and legitimacy to be active. Indeed, because they had been leading a young professional-focused session every year during the Stockholm World Water Week, the WYN and its partners had the legitimacy and the ability to lead a consultation. By confronting their vision of the role of young people in the water sectors, both senior and junior participants have learnt from the current situation and how to move forward. Likewise, by organizing a set of training and workshops on the SDGs dedicated to young people and other stakeholder groups, the WYN was able to gain insights into what specific capacities are missing among its members and among young people in general, and how a better understanding of the SDGs and of the sector can lead to action. Finally, the Kenyan project demonstrates that local and scientific knowledge is complementary if a structured stakeholder engagement and collaborative approach is followed. For knowledge to be co-created and innovations to be accepted and implemented, there is the need to involve the end users and understand their perceptions, concerns and needs. In this project, community members and young scientists converged at a particular time to learn from each other and to share their experiences. The collaborative process allowed community members to better understand the situation regarding the El Niño phenomenon. Moreover, it helped young scientists get insight into the characteristics, specificities, and mental and cultural models of community members regarding disaster-risk management. They could then build on this knowledge and devise actions for the prevention of and responses to disaster.

Changes in behaviours and actions resulting from new understandings

As a result of the projects presented here, and of similar work, the increased number of youth-led fora and sessions organized by the WYN and/with other youth-led organizations at international events demonstrates a change in behaviour of international agencies and institutions. The shift in the set-up of these fora and sessions is also a clear indicator of this change. A few years ago, youth-led sessions focused on 'why' young people needed to participate in international events and global processes. Today, and the session described in this article is an example, youth-led sessions focus mainly on the 'how': what is the role of young people in shaping the OECD water governance principles (Cumiskey, Hoang, Suzuki, Pettigrew, & Herrgård, 2015). They even went a step further. They are already looking for ways that young people can contribute: how can the OECD water governance principles be used for youth-led water operational projects? The change in behaviour can also be observed in the greater number of senior professionals attending youth-led sessions, and the number of sessions being hosted by 'senior-led' agencies and youth-led organizations.

A similar change of behaviours can be observed in the implementation of the SDGs and the Sendai Framework for Disaster Risk Reduction. Since the start of preparation of the 2030 Agenda for Sustainable Development, the number of programmes and financial mechanisms that support 'youth-to-youth' capacity development, co-creation and sharing of knowledge and youth-led initiatives has significantly increased. The projects presented in this article are concrete illustrations of this trend. The ERASMUS+ programme funded by the European Union, by which the YiA4SDG project was supported, is a good example. Likewise, the Youth Science-Policy Interface Platform, led by the United Nations Major Group for Children and Youth, supports the share of knowledge generated by multisector young and early career scientists, engineers, practitioners and policy makers relevant to the science-policy-practice nexus for sustainable development. The disaster-risk reduction project in Kenya is a successful example of the science-practice nexus.

An emergent property of the process to transform the situation

A transformation is also visible in the greater cooperation between young people and different youth-led organizations. Social learning at local and global scales has helped clarify the roles, dependencies, strengths and challenges of different youth-led organizations. This process has facilitated the convergence of positions and finding common points for collaboration. The three example cases related to the WYN in this article serve to demonstrate this transformation. The consultation at the World Water Week was organized by the WYN in collaboration with another youth-led organization, the World Youth Parliament for Water. The YiA4SDG project is the result of a consortia of eight youth-led organizations from Europe and Asia. Finally, the disaster-risk reduction project demonstrates the success of multi-scale collaboration (from global to local) as well as between policy makers (the United Nations Major Group for Children and Youth), young scientists (two WYN and five Kenyan young researchers) and practitioners (local community members).

Discussion

Based on the case-specific results presented above, we have collected 'lessons learned' for each of the elements in the dynamics of stakeholder engagement and social learning. These are summarized in Table 2 and discussed below.

Our findings demonstrate that the distinct modalities for stakeholder engagement explored in these cases require more than a top-down decision to make them work as well as to reap the benefits of social learning.

In line with the OECD principle of stakeholder engagement, the UK and WeSenseIt cases clearly aimed to recognize 'the range of actors with a stake in a situation and

Table 2. Overview of lessons learned from each case.

	UK	Water Youth Network	WeSenseIt
Lessons about 'Who, what, where, when, how and why?'	• Participants were drawn from policy and practitioner communities. The high general level of expertise and familiarity with workshop-type settings meant that a complex inquiry process was possible in the short time available. A more 'public' event would have required more resourcing and time. • The reason for the stakeholder involvement arose from invitations to the researchers to convene the inquiry process. This institutional support and endorsement had the advantage of encouraging other stakeholders to participate. But there is a risk of compromising the independence of the research. • The relatively short time frame of the inquiries reported here means that discussions were inevitably curtailed at times. However, the inquiry design and process did allow the substantive elements to emerge.	• Even a once-a-year event (i.e. Stockholm World Water Week) can aid awareness-raising and capacity development and, as a result, collective action towards sustainable development. • Also serves to enhance collaboration between multi-scale organizations as well as between policy makers, young scientists and practitioners, and as intergenerational dialogue.	• Developing observatories relies on multi-lateral, not just bilateral interactions among stakeholders.
Lessons about the convergence of goals (expressed as purpose)	• The inquiry process provided a structured dialogue on the *purpose* of water governance and identification of core themes for water governance systems. The ability to formulate and agree on a definition of the purpose of current and future water governance indicates convergence of goals at the system level. Convergence does not mean there is always complete agreement, but a recognition of the systemic relationships between different stakeholders' goals, which can then lead to more concerted forms of action.	• The creation of a structured approach for engaging under-represented groups such as young people and local communities is critical for reducing risks, enhancing trust and as a result unlocking their participation in decision-making processes.	• The extent to which convergence of goals can be achieved is highly case-dependent and locally defined.

(*Continued*)

Table 2. (Continued).

	UK	Water Youth Network	WeSenseIt
Lessons about the process of co-creation of knowledge which provides insights into the causes of a situation and the means of its possible transformation	• Process requires an element of trust and willingness to cooperate. • Collaborative process and tasks were challenging in terms of revealing assumptions about the situation, promoting dialogue and integrating different perspectives. • Making boundary choices about the relevance of issues was difficult. • Co-creating and using system diagrams helped reveal the complexity of the situation and the realization that no individual or organization could manage this complexity in isolation. • Water governance is not a 'thing' or single entity that can be added to a situation to bring about improvement. Instead, water governance can be thought of as a learning system.	• Positioning WYN as a trusted youth-led organization has required the organization of successful intergenerational sessions at international events since 2013. • The WYN members had to constantly demonstrate their professionalism.	• Different actors have differing perceptions of flood risk, and citizen observatories can serve to create a more level playing field. • Most interactions still took place offline, not online. • It takes time to build/transfer trust and responsibilities into the online environment, especially since this goes hand in hand with the responsibility to respond.
Lessons about the changes in behaviours and actions	• Changes in behaviours and actions may take time to develop and become 'acceptable' to individuals and organizations. • Not all changes in understanding and practices can be realized by participants. • Evidence of changes is therefore difficult to collect with short-term interventions. However, in the UK case, participants went on to organize other forms of stakeholder involvement into related issues, suggesting that a new form of practice was engendered.	• Changes in behaviours require time and perseverance.	• The changes in behaviour triggered by the observatories took time to materialize (beyond the project life time).

(*Continued*)

Table 2. (Continued).

	UK	Water Youth Network	WeSenseIt
Lessons about the emergent property of the process to transform a situation	• The conditions for social learning can be enabled and facilitated, but there are no guarantees of social learning or any other collaborative element (e.g. convergence of goals). Social learning is neither a magic ingredient nor an assumed outcome.	• Collaboration among organizations working with under-represented groups, including youth-led organizations, is critical for strengthening the position of these groups. • Enhancing trust and collaboration requires time and effort. However, already in the short term, the advantages of such collaboration are noticeable. This multi-scale and multi-sector collaboration can only be achieved with a well-defined social learning process.	• Citizen observatories provide new conditions for social learning but should not be seen as a panacea or as a 'quick fix' for stakeholder engagement. • Demonstrated commitment of all stakeholders leads to trust. • Ultimately, the potential for changing the role of citizens is highly dependent on the room that citizens are granted by authorities – but also on that claimed by citizen.

understanding their possibly diverse responsibilities' (OECD, 2015), while the WYN case is more concerned with ensuring the involvement of under-represented groups whose stakes may have been less evident in various decision-making processes.

Confirming the findings of Furber et al. (2016), the WeSenseIt case showed that stakeholder involvement is not only highly dependent on a sound understanding of stakeholder responsibilities. It also relies on the stakeholders' perceived gains from the collaboration – and is less successful with those stakeholders who perceive the resulting outcomes, such as additional responsibilities, as a loss. Careful attention needs to be paid during the engagement process to how to address individual (perceived) losses and to what extent these may be offset by expected collective gains.

All three cases clearly demonstrate that regardless of the type of stakeholder group involved and the type of intervention chosen, social learning and stakeholder engagement in water governance take considerable time to result in changes in behaviour and actions. Sufficient levels of trust, ownership and continuity are the basis for achieving desired outcomes of social learning. These can be obtained in different ways, though, as illustrated by the well-defined social learning processes of the WYN and the UK cases compared to the exploratory social learning process of WeSenseIt. Nevertheless, generating trust in the ICT-enabled stakeholder engagement of WeSenseIt did encounter additional hurdles.

Moreover, the starting point for social learning appears to be dependent not only on identifying the process of decision making and stakeholder inputs, as indicated by the OECD principles, but also on ensuring that all stakeholders have access to the same information. This in itself is an elaborate process and goes hand in hand with beginning to agree on issue boundaries.

The WYN case presented here shows the dependence of stakeholder engagement on a structured approach in order to include under-represented groups, which requires careful planning of time and resources. This extends beyond water governance specifically to the attainment of the Sustainable Development Goals more generally.

Regarding the contextualization of stakeholder engagement initiatives, our results show that stakeholder engagement, when understood and designed as a social learning process, can lead to shared understanding and concerted actions to improve water governance. Our analysis that stakeholder engagement is *purposeful* – i.e. not a state but an ongoing dynamic – accords well with the emphasis set out in the OECD principle. When 'assessing the process and outcomes of stakeholder engagement to learn, adjust and improve accordingly', it is arguably most salient to focus on how to foster collective learning and cooperation among stakeholders to leverage the potential for sustainable transformation and change. This has implications for rethinking stakeholder engagement in a range of water governance situations to avoid top-down, 'ticking the box' processes which fail to recognize the potential for situation improvement.

Conclusions

In this article, prompted by the OCED principle of stakeholder engagement, we asked a central question: How should we conceptualize and enact stakeholder engagement? Although it is an ex post analysis, based on the findings reported here and consistent with the literature, we find that stakeholder engagement is not just about participation. As Collins and Ison (2009) note, participation is a requirement but not, of itself, sufficient for social learning. Our analysis suggests that a key consideration is to ask: stakeholder engagement *for what purpose*? This is not idle speculation but requires consideration of the ethics, process, participants, roles and expected outcomes of stakeholder engagement, as reflected in the OECD principle.

A reframing of stakeholder engagement as a process of social learning opens up more possibilities than just participation as it carries an explicit purpose which underpins design and process considerations. It also opens up discussion of the responsibilities of those involved as initiators, designers, facilitators, participants and 'recipients' of the process. If no changes are likely, due to, for example, prohibitive institutional arrangements, then inviting stakeholders into a process predicated on social learning and dialogue is ethically questionable. It follows that while designers and facilitators cannot be held to account if the stakeholder engagement 'fails' in terms of social learning, they are responsible for ensuring that the enabling conditions for social learning are met.

In offering insights into a wide range of contexts, levels and processes relating to water governance and stakeholder engagement, our research findings suggest that the term has, at last, begun to be appreciated in a more sophisticated way, with due recognition for the potential for learning. Pitfalls and constraints remain, not least concerns about time, resources, replicability and representation. Nevertheless, stakeholder engagement designed as social learning offers much scope for developing informed and outcome-oriented contributions to water policy design and implementation.

Acknowledgements

The WYN projects were led by several teams and under different sources. The World Water Week consultations received funding support from the Asian Development Bank, IHE Delft and SIWI. We appreciate the guidance from Chris Morris and Ponce Samandiego, and the work of our colleagues: Rozemarijn ter Horst, Cecilia Alda, Veronica Diaz Sosa, Veronica Minaya, Ibrahim Bah, Tlhoriso Morienyane, Bianca Magali Benitez Montiel, Alejandra Molina and Shabana Abbas, as well as the collaboration of the World Youth Parliament for Water. The YiA4SDG project received funding from the European Union (grant agreement 2015-3597). We acknowledge the work of our colleague Shabana Abbas and the collaboration of our partner organizations. Finally, we are grateful for the excellent research undertaken by our colleagues Jack Wachira and Lydia Cumiskey, as well as the support from the local young researchers in the research project in Kenya.

Disclosure statement

No potential conflict of interest was reported by the authors.

Funding

The WeSenseIt case reported in this paper is part of the WeSenseIt project which has received funding from the European Union [Grant no. 308429.ect.] The UK case is informed by the Climate Adaptation and Water Governance Project (http://www.cadwago.net), funded by Riksbankens Jubileumsfond, Compagnia di San Paolo, and VolkswagenStiftung, as part of the Europe and Global Challenges programme [Grant no. GC12-1545:1].

References

Adger, W. N., Paavola, J., Huq, S., & Mace, M. J. (2006). *Fairness in adaptation to climate change*. Cambridge, London: The MIT Press.

Akhmouch, A., & Clavreul, D. (2016). Stakeholder engagement for inclusive water governance:"Practicing What We Preach" with the OECD water governance initiative. *Water*, *8*(5), 204.

Arnstein, S. (1969). A ladder of citizen participation. *J Am Plan Assoc*, *35*(4), 216–224.

Bandura, A. (1977). *Social learning theory*. Englewood Cliffs, NJ: Prentice-Hall.

Barreteau, O., Bots, P. W. G., & Daniell, K. A. (2010). A framework for clarifying "participation" in participatory research to prevent its rejection for the wrong reasons. *Ecology and Society*, *15* (2), 1. doi: 10.5751/ES-03186-150201

Basco-Carrera, L., Warren, A., van Beek, E., Jonoski, A., & Giardino, A. (2017). Collaborative modelling or participatory modelling? A framework for water resources management. *Environmental Modelling & Software*, *91*, 95–110. doi:10.1016/j.envsoft.2017.01.014

Behagel, J., & Turnhout, E. (2011). Democratic legitimacy in the implementation of the water framework directive in the Netherlands: Towards participatory and deliberative norms? *Journal of Environmental Policy & Planning*, *13*(3), 297–316. doi:10.1080/1523908X.2011.607002

Blackmore, C., Ison, R., & Jiggins, J. (2007). Social learning: An alternative policy instrument for managing in the context of Europe's water. *Environmental Science & Policy*, *10*(6), 493–498. doi:10.1016/j.envsci.2007.04.003

Bourget, L. (Ed.). (2011). *Converging waters: Integrating collaborative modeling with participatory processes to make water resources decisions*. Washington DC: Institute for Water Resources, U.S. Army Corps of Engineers.

Bruns, B. (2003) Water Tenure Reform: Developing an Extended Ladder of Participation, paper presented at the 'Politics of the Commons: Articulating Development and Strengthening Local Practices' conference, July 11-14, 2003, Chiang Mai, Thailand

Checkland, P. (1981). *Systems thinking, systems practice*. Chichester: John Wiley.

Checkland, P., & Scholes, J. (2002). *Soft Systems methodology in action*. Chichester: John Wiley.

Cleaver, F. (1999). Paradoxes of participation: Questioning participatory approaches to development. *Journal of International Development*, *11*(4), 597–612.

Cockerill, V. C., Tidwell, V. C., Passell, H. D., & Malczynsky, L. A. (2007). Cooperative modelling lessons for environmental management. *Environmental Practice*, *9*(1), 28–41. doi:10.1017/S1466046607070032

Collins, K., Blackmore, C., Morris, R., & Watson, D. (2007). A systemic approach to managing multiple perspectives and stakeholding in water catchments: Some findings from three UK case studies. *Environmental Science and Policy*, *10*(6), 564–574. doi:10.1016/j.envsci.2006.12.005

Collins, K., & Ison, R. (2009). Jumping off Arnstein's Ladder: Social learning as a new policy paradigm for climate change adaptation. *Environmental Policy and Governance*, *19*(6), 358–373. doi:10.1002/eet.v19:6

Colvin, J., Blackmore, C., Chimbuya, S., Collins, K., Dent, M., Goss, J., . . . Seddaiu, G. (2014). In search of systemic innovation for sustainable development: A design praxis emerging from a decade of social learning inquiry. *Research Policy*, *43*, 760–771. doi:10.1016/j.respol.2013.12.010

Cumiskey, L., Hoang, T., Suzuki, S., Pettigrew, C., & Herrgård, M. M. (2015). Youth participation at the Third UN World Conference on disaster risk reduction. *International Journal of Disaster Risk Science*, *6*(2), 150–163. doi:10.1007/s13753-015-0054-5

DeSario, J., & Langton, S. (1987). Citizen participation and technocracy. In L. DeSario & S. Langton (Eds.), *Citizen participation in public decision making*. New York: Greenwood Press.

Edelenbos, J., & Klijn, E.-H. (2006). Managing stakeholder involvement in decision-making: A comparative analysis of six interactive processes in the Netherlands.. *Journal of Public Administration Research and Theory*, *16*(3), 417–446. doi:10.1093/jopart/mui049

Edelenbos, J., Van Schie, N., & Gerrits, L. (2008). Democratic anchorage of interactive governance: Developing institutional interfaces in water governance. In Proceedings 2008 conference of the Political Science Association. http://www.psa.uk/proceedings.aspx.

Evers, M., Jonoski, A., Maksimovič, Č., Lange, L., Ochoa Rodriguez, S., Teklesadik, A., … Makropoulos, C. (2012). Collaborative modelling for active involvement of stakeholders in urban flood risk management. *Natural Hazards and Earth System Science, 12*(9), 2821–2842. doi:10.5194/nhess-12-2821-2012

Fischer, F. (2000). *Citizens, experts, and the environment: The politics of local knowledge.* Durham: Duke University Press.

Foster, N., Collins, K., Ison, R., & Blackmore, C. (2016). Water Governance in England: Improving understandings and practices through systemic co-inquiry. *Water, 8*, 540–556. doi:10.3390/w8110540

Fung, A. (2006). Varieties of participation in complex governance. *Public Administration Review, 66*, 66–75. doi:10.1111/puar.2006.66.issue-s1

Furber, A., Medema, W., Adamowski, J., Clamen, M., & Vijay, M. (2016). Conflict management in participatory approaches to water management: A case study of lake ontario and the St. Lawrence River Regulation. *Water 2016, 8*(7), 280–296.

Greenwood, D. J., & Levin, M. (1998). *Introduction to action research: Social research for social change.* Thousand Oaks, CA: Sage Publications.

Gunderson, L. H. (2003). Adaptive dancing: Interactions between social resilience and ecological crises. In F. Berkes, J. Colding, & C. Folke (Eds.), *Navigating social-ecological systems. Building resilience for complexity and change* (pp. 33–52). Cambridge: Cambridge University Press.

GWP (2000). Integrated water resources management, TAC Background Papers. Author.

Hare, M. (2011). Forms of participatory modelling and its potential for widespread adoption in the water sector. *Environmental Policy and Governance, 21*, 386–402. doi:10.1002/eet.590

Hare, M., Letcher, R. A., & Jakeman, A. J. (2003). Participatory modelling in natural resource management: A comparison of four case studies. *Integrated Assessment, 4*(2), 62–72. doi:10.1076/iaij.4.2.62.16706

Hisschemöller, M. (1993). *De Democratie van problemen, de relatie tussen de inhoud van beleidsproblement en methoden van politieke besluitvorming.* Amsterdam: VU-Uitgeverij.

Hurlbert, M., & Gupta, J. (2015). The split ladder of participation: A diagnostic, strategic, and evaluation tool to assess when participation is necessary. *Environmental Science & Policy, 50*, 100–113. doi:10.1016/j.envsci.2015.01.011

Irvin, R. A., & Stansbury, J. (2004). Citizen participation in decision making: Is it worth the effort? *Public Administration Review, 64*(1), 55–65. doi:10.1111/puar.2004.64.issue-1

Ison, R. (2010). Systemic inquiry. In *Systems practice: How to act in a climate-change world* (pp. 243–265). London: Springer.

Ison, R., Röling, N., & Watson, D. (2007). Challenges to science and society in the sustainable management and use of water: Investigating the role of social learning. *Environmental Science & Policy, 10*(6), 499–511. doi:10.1016/j.envsci.2007.02.008

Ison, R. L., Collins, K., & Wallis, P. (2015). Institutionalising social learning: Towards systemic and adaptive governance. *Environmental Science & Policy, 53*(Part B), 105–117. doi:10.1016/j.envsci.2014.11.002

Jonoski, A., & Evers, M. (2013). Sociotechnical framework for participatory flood risk management via collaborative modeling. *International Journal of Information Systems and Social Change (IJISSC), 4*(2), 1–16. doi:10.4018/IJISSC

Joshi, S., & Wehn, U. (2017). From assumptions to artifacts: Unfolding e-participation within multi-level governance. *Electronic Journal of e-Government, 15*(2), 116–129.

Lewin, K. (1946). Action research and minority problems. *Journal of Social Issues, 2*(4), 34–46. doi:10.1111/josi.1946.2.issue-4

Loucks, D. P., Van Beek, E., Stedinger, J. R., Dijkman, J. P., & Villars, M. T. (2005). *Water resources systems planning and management: An introduction to methods, models and applications.* Paris: UNESCO.

Mayer, I. S., van Bueren, E. M., & Bots, P. (2005). Collaborative decision making for sustainable urban renewal projects: a simulation-gaming approach. *Environment and Planning B: Urban Analytics and City Science, 32*(3), 403–423.

Mostert, E., Pahl-Wostl, C., Rees, Y., Searle, B., Tàbara, D., & Tippett, J. (2007). Social learning in European river-basin management: Barriers and fostering mechanisms from 10 river basins. *Ecology and Society, 12*(1), art. 19. doi:10.5751/ES-01960-120119

OECD. (2015), Stakeholder engagement for inclusive water governance, OECD Studies on Water, Author, Paris.

Pahl-Wostl, C., & Hare, M. (2004). Processes of social learning in integrated resources management. *Journal of Community & Applied Social Psychology, 14*(3), 193–206. doi:10.1002/(ISSN)1099-1298

Pahl-Wostl, C., Tàbara, D., Bouwen, R., Craps, M., Dewulf, A., Mostert, E., ... Taillieu, T. (2008). The importance of social learning and culture for sustainable water management. *Ecological Economics, 64*(3), 484–495. doi:10.1016/j.ecolecon.2007.08.007

Reed, M. (2008). Stakeholder participation for environmental management: A literature review. *Biological Conservation, 141*(10), 2417–2431. doi:10.1016/j.biocon.2008.07.014

Rinaudo, J. D., & Garin, P. (2005). The benefits of combining lay and expert input for water-management planning at the watershed level. *Water Policy, 7*(3), 279–293.

Röling, N. (2002). Beyond the aggregation of individual preferences. In C. Leeuwis & R. Pyburn (Eds.), *Wheelbarrows full of frogs. Social learning in rural resource management* (pp. 25–47). Aasen: Koninklijke Van Gorcum.

Rowe, G., & Frewer, L. (2004). Evaluating public participation exercises: A research agenda. *Science, Technology, & Human Values, 29*(4), 512–556. doi:10.1177/0162243903259197

Sadoff, C. W., & Grey, D. (2005). Cooperation on international rivers: A continuum for securing and sharing benefits. *Water International, 30*(4), 420–427. doi:10.1080/02508060508691886

Scholz, G., Dewulf, A., & Pahl-Wostl, C. (2014). An analytical framework of social learning facilitated by participatory methods. *Journal of Systemic Practice and Action Research, 27*(6), 575–591. doi:10.1007/s11213-013-9310-z

SLIM. 2004a. Stakeholders and stakeholding in integrated catchment management and sustainable use of water. SLIM Policy Brief No.2. SLIM, UK.

SLIM. 2004b. SLIM Framework: Social Learning as a Policy Approach for Sustainable Use of Water, SLIM.

Sørenson, E. (2002). Democratic theory and network governance. *Administrative Theory and Praxis, 24*(4), 693–720.

Sørenson, E., & Torfing, J. (Eds.). (2007). *Theories of democratic network governance.* New York: Palgrave Macmillan.

van Buuren, A., Driessen, P., Teisman, G., & van Rijswick, M. (2014). Toward legitimate governance strategies for climate adaptation in the Netherlands: Combining insights from a legal, planning, and network perspective. *Regional Environmental Change, 14*, 1021–1033.

Voinov, A., & Bousquet, F. (2010, November). Modelling with stakeholders. *Environmental Modelling & Software, 25*(11), 1268–1281. doi:10.1016/j.envsoft.2010.03.007

Voinov, A., Kolagani, N., McCall, M. K., Glynn, P. D., Kragt, M. E., Ostermann, F. O., ... Ramu, P. (2016, March). Modelling with stakeholders – Next generation. *Environmental Modelling & Software, 77*, 196–220. doi:10.1016/j.envsoft.2015.11.016

von Korff, Y., Daniell, K. A., Moellenkamp, S., Bots, P., & Bijlsma, R. M. (2012). Implementing participatory water management: Recent advances in theory, practice, and evaluation. *Ecology and Society, 17*(1), art. 30. doi:10.5751/ES-04733-170130

Watson, N. (2014). IWRM in England: Bridging the gap between top-down and bottom-up implementation. *International Journal of Water Resources Development, 30*(3), 445–459. doi:10.1080/07900627.2014.899892

Wehn, U., & Evers, J. (2015). The social innovation potential of ICT-enabled citizen observatories to increase eParticipation in local flood risk management. *Technology in Society, 42*, 187–198. doi:10.1016/j.techsoc.2015.05.002

Wehn, U., McCarty, S., Lanfranchi, V., & Tapsell, S. (2015). Citizen observatories as facilitators of change in water governance? Experiences from three European cases, Special Issue on ICTs and Water. *Journal of Environmental Engineering and Management, 14*(9), 2073–2086.

Wehn, U., Rusca, M., Evers, J., & Lanfranchi, V. (2015). Participation in flood risk management and the potential of citizen observatories: A governance analysis. *Environmental Science and Policy, 48*(April), 225–236. doi:10.1016/j.envsci.2014.12.017

Wendling, C., Radisch, J., & Jacobzone, S. (2013). The use of social media in risk and crisis communication. OECD Working Papers on Public Governance, No. 24, OECD Publishing, Paris.

Woodhill, J., & Röling, N. (1998). The second wing of the eagle: The human dimension in learning our way to more sustainable futures. In N. G. Roling & M. A. E. Wagemakers (eds), *Facilitating sustainable agriculture. Participatory learning and adaptive management in times of environmental uncertainty* (pp. 46–71). Cambridge: Cambridge University Press.

Zeitoun, M., & Mirumachi, N. (2008). Transboundary water interaction I: Reconsidering conflict and cooperation. International environmental agreements: Politics. *Law and Economics, 8*(4), 297–316.

OECD Principles on Water Governance in practice: an assessment of existing frameworks in Europe, Asia-Pacific, Africa and South America

Susana Neto, Jeff Camkin, Andrew Fenemor, Poh-Ling Tan, Jaime Melo Baptista, Marcia Ribeiro, Roland Schulze, Sabine Stuart-Hill, Chris Spray and Rahmah Elfithri

ABSTRACT

Through the lens of the 12 OECD Principles on Water Governance, this article examines six water resources and water services frameworks in Europe, Asia-Pacific, Africa and South America to understand enhancing and constraining contextual factors. Qualitative and quantitative methods are used to analyze each framework against four criteria: alignment; implementation; on-ground results; and policy impact. Four main target areas are identified for improving water governance: policy coherence; financing; managing trade-offs; and ensuring integrity and transparency by all decision makers and stakeholders. Suggestions are presented to support practical implementation of the principles through better government action and stakeholder involvement.

Introduction and objectives

Objectives of this article and expected results

This article explores the extent to which a selection of existing water governance frameworks align with the 12 Principles on Water Governance, released by the Organisation for Economic Co-operation and Development (OECD) in 2015. The article identifies factors that enable or constrain the progress of different water governance approaches in different regions of the world, by analyzing how well a set of existing frameworks provide conditions for effective implementation of the OECD principles. This analysis of what influences the implementation of the OECD principles provides a basis for further recommendations, in relation to both water resources and water services.

We examine six frameworks with different characteristics, chosen in order to consider the OECD principles in a wide range of different contexts. We include four national water policy frameworks (Australia, Brazil, New Zealand and South Africa), one transnational water policy framework (Europe) and one global guideline (Lisbon Charter, using the example of Portugal), each with a different focus. For example, the EU Water Framework Directive (WFD) was established with the main objectives of building a comprehensive legal framework for water quality and re-establishing ecological integrity; the main driver of South Africa's water policy was initially to rebalance access to water in the post-apartheid period; and a higher aim of Brazil's water policy has been to democratize water management. While all the frameworks address the three main drivers of water governance (effectiveness, efficiency, and trust and engagement), some weightings are evident. For example, the New Zealand and South African water policy frameworks, and the Lisbon Charter, are weighted towards regulatory effectiveness, the WFD and Australia's National Water Initiative (NWI) towards increasing efficiency, and Brazil's Water Resource Management Policy towards building trust and engagement. Lessons from the UNESCO-IHP Hydrology for the Environment, Life and Policy programme (UNESCO, 2010), which has been influential in some of the countries analyzed, were implicit in the assessments and conclusions drawn.

As the OECD principles refer to water resources and water services, it is useful to note that the European WFD and the cases of Australia, New Zealand and Brazil are more focused on water resources, the case of South Africa is focused both on water resources and water services, and the Lisbon Charter (using the example of Portugal) is more focused on water services. Our main aim was not to compare the frameworks with each other, but to compare their 'performance' against the 12 OECD principles. We designed criteria specifically to obtain a clearer diagnostic of the synergies, opportunities and constraints on implementation for each framework. Based on this, we make recommendations for more effective implementation through better context-focused governmental action and stakeholder involvement in relation to the OECD principles.

Background to the OECD Principles on Water Governance

The concept of 'water governance' has gained prominence in recent years due to concerns over water as a societal risk, triggered by increased competition of use in a context of change (Woodhouse & Muller, 2017). Scholars have canvassed diverse views on water governance, ranging from institutional analysis (Jager et al., 2016; Neto, 2010), to adapting managerial theories to environmental learning (Pahl-Wostl, 2009), and leadership (Meijerink & Huitema, 2010; Taylor, Wouter, Arriëns, & Laing, 2015). Some scholars question meso-scale policies while advocating global evaluation indicators (Biswas & Tortajada, 2010); others call for context-specific criteria in water governance (Akmouch & Correia, 2016).

This article adopts the practitioner-oriented definition of water governance developed by the OECD (2015a): 'Water governance is the set of rules, practices, and processes (formal and informal) through which decisions for the management of water resources and services are taken and implemented, stakeholders articulate their interest and decision-makers are held accountable.'

Akmouch and Correia (2016) reviewed the work done by OECD in recent years on water governance and discussed the development and application of the Principles on Water Governance (Figure 1). The principles were developed by the OECD Water Governance Initiative, a multi-stakeholder platform of over 100 delegates from public, private and non-profit sectors, to support collective action to scale up governance responses to water challenges. The 12 principles are presented in Appendix 1.

The principles are clustered around three main drivers – effectiveness, efficiency, and the ability to generate trust and engagement – defined as follows by the OECD (2015a):

> *Effectiveness* relates to the contribution of governance to define clear sustainable water policy goals and targets at all levels of government, to implement those policy goals, and to meet expectation targets.
>
> *Efficiency* relates to the contribution of governance to maximise the benefits of sustainable water management and welfare at the least cost to society.
>
> *Trust and engagement* relates to the contribution of governance to building public confidence and ensuring inclusiveness of stakeholders through democratic legitimacy and fairness for society at large.

Since adoption, 42 countries and more than 140 major stakeholder groups have endorsed the principles. Work is now underway through the Water Governance Initiative to develop indicators and assess good water governance practices.

Figure 1. Overview of the OECD Principles on Water Governance (OECD, 2015a).

Assessment methodology

Six frameworks were included in the analysis, selected on the basis that the authors had strong familiarity with them and/or sufficient published information was available. Each framework was considered in relation to the 12 OECD principles using four criteria: alignment; implementation; on-ground results; and policy impacts. A Likert scale from 1 to 5 was used (Table 1), and the descriptors used to guide the assessments are fully detailed in Appendix 2. Assessments apply to the present time, while recognising that the last three criteria may be works in progress.

The author responsible for each case study used public information and expert opinion to identify patterns that could increase our understanding of what factors might enhance or constrain implementation of the OECD principles. All of the authors contributed to drawing out common messages that might explain the patterns, to draw conclusions and make suggestions on how water governance can be improved in practice.

Presenting the frameworks of water governance

European Water Framework Directive

Facing the inadequacy of the European Community legislation relating to the protection of waters then currently in force, the European Commission presented a draft directive (European Commission, 1993) on the ecological quality of waters in 1994. A real change of paradigm was evident, from the earlier rationale of public health and the preservation of competition conditions in the common market to a global vision of the environment which Europe pledged to protect and enhance to ensure its long-term sustainable development (Correia, 2003). Sustainability was the keyword for this paradigm shift. This new framework for action, approved for Europe in 2000, brought into management a new perspective which was more comprehensive and integrated in qualitative and quantitative terms, as well as addressing ecological and socio-economic aspects.

The WFD precipitated a fundamental change in management objectives, from merely pollution control to ensuring ecosystem integrity as a whole (Hering et al., 2010). However, as noted by Jager et al. (2016), this has been accompanied by a 'diversity of institutional approaches', and 'overall, the WFD has driven a highly uneven shift to river basin-level planning among member states', while at the same time instigating 'a range of efforts to institutionalise stakeholder involvement'.

The development framework of the management policies of the WFD is set at the scale of river basins defined by topographical boundaries, regardless of the territorial and administrative boundaries of the member states. The WFD defines a river basin

Table 1. Adapted Likert scale for scoring each policy framework.

Score	Alignment	Implementation	On-ground results	Policy impact
1	No alignment	No implementation	No evidence of change	No impact
2	Poor	Poor	Poor	Poor
3	Moderate	Moderate	Moderate	Moderate
4	Good/strong	Good/strong	Good/strong	Good/strong
5	Full alignment	Full implementation	Major change evident	Very strong impact

district, which is the main management unit for river basins, as 'the area of land and sea constituted by one or more neighbouring river basins and by underground and coastal waters associated to them'. The environmental objectives for the different types and categories of water bodies mandate that each member state conduct a characterization of the river basin district, an environmental impact analysis of human activity and an economic analysis of water use. The status of implementation of the WFD through the river basin planning process, updated for 2016, is presented in Figure 2.

Arguably, the WFD was a 'top-down' approach to catchment management, with objectives and instruments set through formal legislative and regulatory processes, and steered by higher governmental organizational levels. Even if the involvement of stakeholders is promoted, the planning process is geared to support the achievement of WFD objectives rather than to respond to broader societal demands (Blackstock, Martin-Ortega, & Spray, 2015; Hendry, 2014; Rouillard & Spray, 2016). But it is undeniable that this framework brought a new comprehensive approach to water management, and one that attempts to integrate it with the ecological, social and economic dimensions (Spray & Comins, 2011). While reaction to the tensions of top-down and bottom-up approaches remains, for example in England and Wales, the

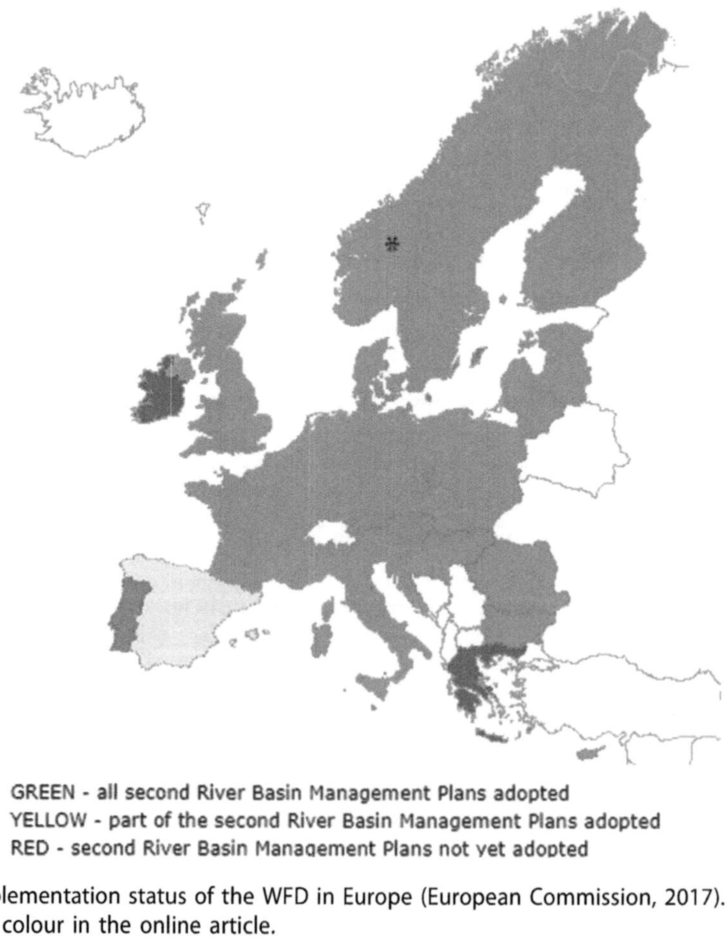

Figure 2. Implementation status of the WFD in Europe (European Commission, 2017). This map can be viewed in colour in the online article.

recent rise of the catchment-based approach, led by NGOs and others in partnership with agencies and other statutory bodies, has been a welcome new development.[1]

Australian National Water Initiative

Australia has an infamous boom-and-bust water economy, cycling through drought and floods. In the early years of colonization by the British this produced a developmental paradigm supporting irrigation in the inland parts of the continent (Connell, Robins, & Dovers, 2007; Lloyd, 1988; Pigram, 2006; Powell, 1989). However, intensive use left an 'indelible mark on Australia's river systems' and the natural environment (Pigram, 2006, p. 61).

Australia's first national water reform was enacted in 1994 to address flaws in the regulatory framework (Hussey & Dovers, 2007; Tan, 2002). The administrative allocation of water had systemic problems stemming from the adoption of English common law concepts of water (Tan, 2002), and in a development era, bureaucrats used their powers to promote water use and politicians to promote their power base (Chenoweth, 1998). Consideration of ecosystems in water management was almost non-existent until a blue-green algal bloom along the Darling River in 1992 made it obvious to politicians and the community alike that the crisis could no longer be ignored.

Under the Australian Constitution, state and territory governments have primary authority over water, and thus national reform can only be undertaken collaboratively through the Council of Australian Governments (Kildea & Williams, 2010). Between 2004 and 2005 all Australian states signed the NWI, placing primary responsibility for implementing the agreement's reform agenda with the state and territory governments, with support from the Australian government.[2]

Commentators such as Connell et al. (2007) note that the NWI's primary goal is sustainable management of water, as seen in the triple-bottom-line statement in Figure 3, yet most of the measures in the NWI relate to water entitlements. The establishment of nationally compatible water access entitlements to facilitate the operation of water markets within and between jurisdictions required changes in water legislation and implementation through state-based statutory water planning processes incorporating public participation of varying degrees (Gentle & Olszak, 2007; Tan, Bowmer, & Mackenzie, 2012).

Driven by the NWI, other institutional reforms were undertaken through the Water Act 2007, which set up the Commonwealth Environmental Water Holder and the Murray-Darling Basin Authority, both with statutory obligations. Community engagement over the 2012 Murray-Darling Basin Plan was a key issue (Evans & Pratchett, 2013; Tan & Auty, 2017). Indigenous interests were acknowledged for the first time in water policy in the NWI (Tan & Jackson, 2013). This has led to stronger requirements in the Water Act 2007 for water plans in the Murray-Darling Basin to articulate and provide for Indigenous values in water.

A statutory watchdog – the National Water Commission (NWC) – also played a critical role in providing guidance for policy making and assessed implementation of water governance and planning. Before the NWC was abolished in early 2015 it carried out four assessments of implementation of the NWI, with the final assessment covering 2004 to 2014 (NWC, 2014a), together with separate report cards on environmental management of water (e.g., NWC, 2010) and the implementation of water plans across

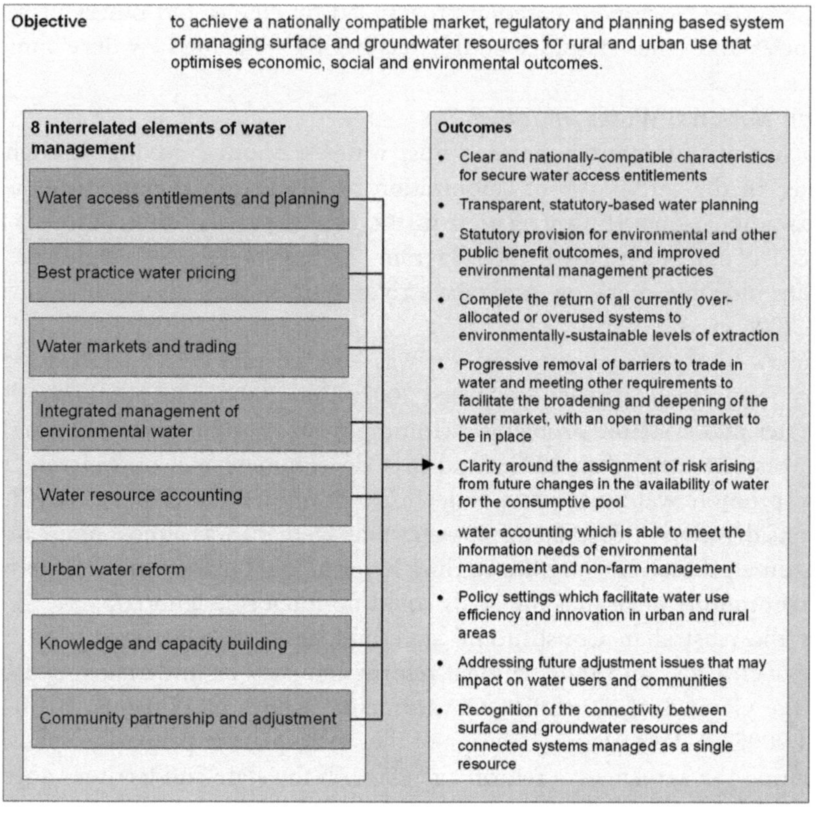

Figure 3. Objective, elements and outcomes of Australia's National Water Initiative (NWC, 2014a).

eight criteria (NWC, 2014b). The NWC's in-depth 2011 assessment of water governance found that the NWI had delivered significant benefits by providing direction for reform but also noted that the major goals of sustainable management and efficiency had yet to be met. The Productivity Commission will now assess water reform and is expected to recommend future reform priorities in late 2017.

New Zealand water policy

Management of freshwater in New Zealand's rivers, lakes and aquifers is governed under its 1991 Resource Management Act, the purpose of which is 'to promote the sustainable management of natural and physical resources'. New Zealand has one of the most devolved water governance regimes in the world (Fenemor, Davie, & Markham, 2006) and one of the first to give legal recognition to Indigenous rights to water, for example in the Waikato and Whanganui catchments (Ruru, 2013; Salmond, 2014).[3] Sixteen regional authorities, whose boundaries mainly follow catchment boundaries (Figure 4), are governed by locally elected councillors and have primary responsibilities that include management of water allocations, water quality management, and flood management.

Figure 4. New Zealand regional councils.

Catchment-based river management began in New Zealand with the 1941 Soil Conservation and Rivers Control Act, which set up catchment boards, predecessors to today's regional councils. The 1991 Resource Management Act largely carried over the water management regime from the 1967 Water and Soil Conservation Act, but introduced a statutory planning regime allowing councils through a hearing process to develop and enforce legally binding objectives, policies and rules at a relevant catchment or regional scale ('regional plans').

Water is managed as a public good. Regional plans are commonly developed at large catchment scale, and increasingly regulate not just water use but also land uses such as intensive dairy farming which are increasing diffuse pollution of water bodies (Duncan, 2014; Monaghan, De Klein, & Muirhead, 2008). The New Zealand government was criticized for its lack of national support to guide implementation during the first 15 years of the Resource Management Act, including the lack of national policy statements or national environmental standards (Jackson & Dixon, 2007). This situation is now being improved. Since 2010, a national Land and Water Forum has used a collaborative approach to recommend methods for setting and managing water within catchment limits (e.g., Land and Water Forum, 2012). Collaborative approaches to catchment planning are now being implemented throughout New Zealand (Fenemor, Neilan, Allen, & Russell, 2011; Rouse & Norton, 2016), with many plans in their second decade of implementation, and giving greater weight to Māori cultural values (Harmsworth, Young, Walker, Clapcott, & James, 2011). Public concerns over declining water quality in lowland rivers, Māori calls for a greater role in water governance, and calls for resource rentals to be paid, especially for water exports, have raised water governance to a national political issue (Fenemor, 2017).

Brazilian national water policy

The National Water Resource Management System (SINGREH) of Brazil was created in 1997 through the Federal Water Law 9433. This law represented the turning point for water management in Brazil by replacing the former Water Act of 1934, defining its policy (principles, objectives, management instruments and institutional framework) and driving the country to a water reform.

The 1997 Water Law addresses six principles for water management: water is a public good; water is a limited resource with economic value; the priority is for human consumption in water-shortage periods; water management must promote multiple uses of water; water is to be managed at the river-basin level; and water resource management should be decentralized and consider participation of users, communities and government. Five management instruments were designed to implement these principles: water plans; water permits; a system for classification of water bodies according to their water quality; bulk water fees; and an information system for water resources.

There are three scales for water planning in Brazil: national, state (each state has its own water law and plan, which should consider national guidelines), and basin. At the basin scale, two water management domains exist: national/federal and state. When a river crosses more than one state and one of its tributary basins is totally within a state area, this sub-basin is subject to state domain. But there are reservoirs built by the federal government, where the stored water is in federal domain, even in basins under state control. Federal and state domains relate to surface water only. The state domain is always applied in groundwater management under a determination made by the 1988 Federal Constitution.

SINGREH is responsible for operating the National Water Resources Policy.[4] At the national level it includes the National Water Resource Council (responsible for elaborating national guidelines), the National Water Agency or ANA (a federal agency for implementing water management at the national level), and federal river basin committees (participatory and decentralized management bodies). At the state level, corresponding entities apply. The river basin committees are consultative, normative and deliberative bodies, and composed of elected representatives of civil society and of water users and appointed government representatives. Water conflicts are to be arbitrated in the first instance at the basin committee. These bodies are intended to implement one of the main principles stated by the 1997 Federal Water Law: a decentralized and participatory approach to guide bottom-up water policy making. However, during the last 20 years, it is observed that the government has had difficulty sharing power (at federal and state levels).

South African national water policy

Water has been an integral component of the political realm in South Africa since the 1930s, when, following the Great Depression, major irrigation schemes were set up to

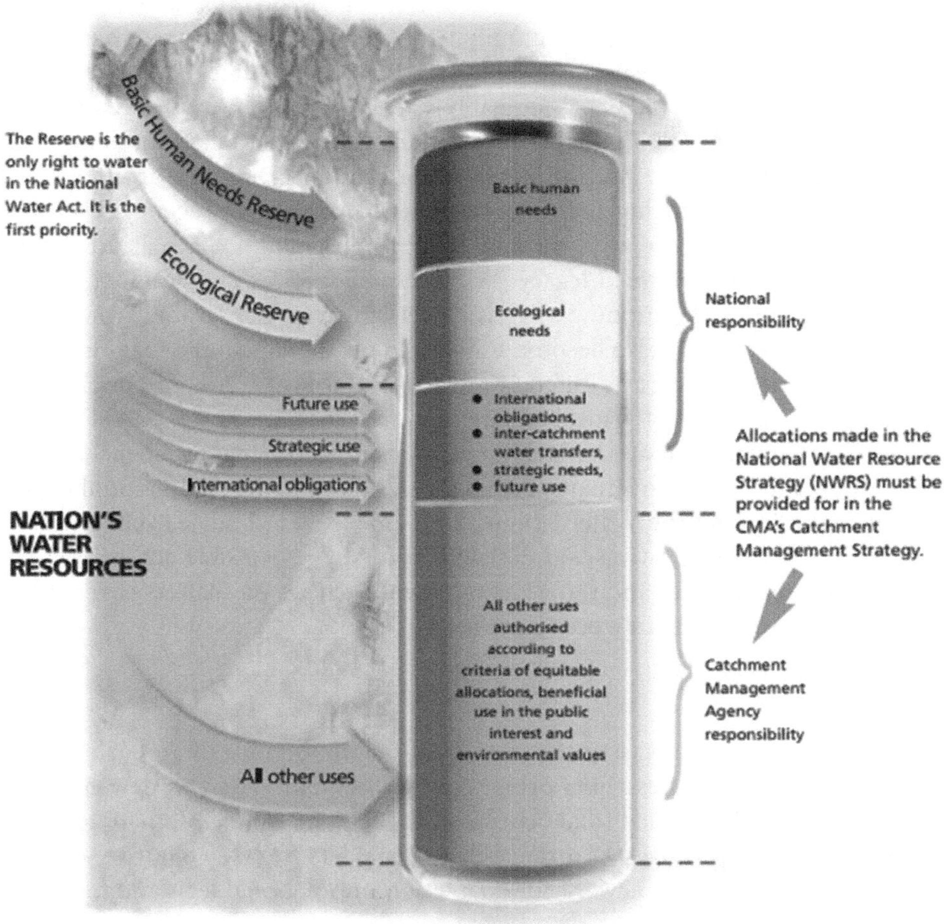

Figure 5. Concepts encapsulated by the National Water Act of South Africa (De la Harpe & Ramsden, 1998).

employ so-called poor whites. From the 1950s, water represented one of the technological flagships for the apartheid government at the time, with engineered water systems dominating the legal, policy and decision-making landscape. Following the advent of a democratic government in South Africa in 1994, the 1956 Act was repealed and replaced by the Water Services Act of 1997 and the National Water Act of 1998. The National Water Act deals with the management of water resources as a constitutional responsibility of the national government (Figure 5).

The Water Services Act also deals with the provision of water supply and sanitation services, which are the constitutional responsibility of local government. It is important to note that these two acts have been rewritten completely and follow, in most areas, principles diametrically opposed to those of the old South African water legislation. Water resources are allocated via a system of free use (uncharged), general authorization and licensing. Water for the Reserve (i.e., the Human Health Reserve and the Ecological Reserve), as well as Strategic Uses, is allocated before that for any other users.

The National Water Resource Strategy (2004, updated in 2013) and the various Catchment Management Strategies guide implementation of the National Water Act.

While the aims of both the National Water Act and the Water Services Act were decentralization of water management, only two catchment management agencies and a handful of water user associations have been established to date.

In the past five years there has arguably been a move towards recentralization, demonstrated by, *inter alia*, a reduction from the original 19 water management areas to 9, and retention of water licensing within the National Department of Water and Sanitation rather than moving it to the catchment management agency level as was originally intended. In many ways the constitution (providing for a right to water) on the one hand, and the apartheid legacy (lack of water services and access to water for the masses) on the other, are now seen as drivers in the water sector. Water has not only been politicized, it has also become a highly emotive public topic, leading to unrest and riots on an annual basis in different areas of the country. It is important to note that while the main focus of water planning and management in the past decades has been on water quantity issues, water quality is an increasing problem. While the framework for good water governance in South Africa exists, the realities on the ground do not match it. Indeed, the Water Tribunal, an independent body established in 1998 to hear appeals against directives and decisions made by responsible authorities, catchment management agencies and water management agencies on matters covered by the National Water Act, has hardly ever been used.

Lisbon Charter

Internationally there is increasing recognition of the importance of an enabling environment for the delivery of essential drinking water, sanitation and wastewater services through good public policy and effective regulation. The International Water Association has brought together a range of stakeholders to take collective action and provide an international framework for adaptation into national legislation, regulation and practices.

Motivated by the challenges faced by many countries in improving the water supply and sanitation situation, and the chasm between willingness to change and real change in the field, in January 2015 the International Water Association adopted the Lisbon Charter to guide public policy on and regulation of drinking water supply, sanitation and wastewater management services around the world (IWA, 2015).[5]

The Lisbon Charter relies on the following fundamental elements: water services are fundamental to the health of communities; they are essential to the sustainable socio-economic development of society; they have been recognized as human rights; governments must ensure water services provision in compliance with their international commitments as well as commitments to their own people; with that goal, governments should foster good public policy and effective regulation.

The charter endorses responsibilities for governments and public administration regarding adoption of strategic plans, strengthening legal frameworks and designing regulatory frameworks. Among other things, it advocates realistic and measurable targets, reliable information, diverse financial and fiscal measures, and in general improvement of the structural efficiency of services to develop the water sector and promote awareness and participation of users. It also includes recommendations about the responsibilities of regulatory authorities, service providers and users. Finally, the charter recommends, as common responsibilities, that all stakeholders have an ongoing and open dialogue, and share information.

The Lisbon Charter principles and the OECD principles are fundamentally matched. In this analysis, to ensure an operational focus, we have used the reform of water services in Portugal, which complies with the Lisbon Charter principles, as an illustration and demonstration of that strong match. The Portuguese implementation of water services is notable, not only regarding the accessibility and quality of drinking water supply and wastewater management, but also with respect to impacts on environmental quality and public health. Considering the integrated approach to water resources and water services that underlines the OECD principles, this case is a good illustration of the benefits the OECD Principles on Water Governance can bring to water services as well.

Assessment and analysis

Results and findings of each framework assessment

The results of the assessment (scoring 1 to 5) of each water policy framework against the four criteria (alignment, implementation, on-ground results, and policy impact) for each of the 12 OECD principles are shown in Appendix 3. Key findings for each policy framework, drawn from those results, are presented below.

European Water Framework Directive

The principles and guidelines of the WFD are, in general, well aligned with the OECD principles. There are some specific approaches (e.g., Principle 6, on water finance, and 11, on trade-offs) where there are different policy targets and spheres of activity, and the alignment scores lower.

The WFD had a strong impact in terms of imposing new regulatory and legislative frameworks in member states, and it also enhanced more integrated approaches between environmental, social and economic dimensions. This triggered several legislative and institutional reforms, so it generally scores high for policy impact.

The criteria related to implementation and on-ground results were assessed at the European level. This assessment of the WFD against the OECD principles covers the European territory and so includes a great diversity of specific national contexts and frameworks of implementation, and consequently diverse internal policy impacts. Within this first overall assessment that diversity cannot be reflected.

Australian National Water Initiative

In general Australia's NWI presents a strong alignment with the OECD principles. Regarding effectiveness, there is evidence of the implementation of the NWI being based on adequate resourcing and an independent 'watchdog'. Rigorous participatory assessments of implementation across all sectors provide transparency and accountability. Timetables for the NWI were necessary but underplayed the complexity of the reform. Integration across water quality and quantity is not yet built into the system of decision making.

The efficiency of the Australian framework relies on good data and information being shared across all jurisdictions. While financing has been adequate, cost-benefit analysis has not always been carried out well. While institutional reform has generally separated the policy, regulation and service delivery functions, urban water governance is still lagging behind in policy development. Pricing of water remains an issue and does not reflect the full cost of delivery. Recognition of Indigenous interests in water only came about through national leadership in policy development and research, with reform translating into specific provisions in state legislation, for example in New South Wales (Tan & Jackson, 2013). Further state actions intend to include Aboriginal values and traditional ecological knowledge in water planning (State of Victoria, 2016). Yet there is much to be done to build trust and engagement, including more transparent trade-offs. Reduction of 'red tape', done in the name of efficiency, has undermined trust in state government processes.

New Zealand water policy

New Zealand's Resource Management Act and catchment-based management since the 1940s created a strong basis for governance devolved to catchment and regional communities. This creates high levels of alignment with the OECD principles.

On-ground results and policy impact scored lower than alignment and implementation for most principles, partly because implementation of catchment-based water management plans containing water take and quality limits is in the early stages in some regions, and because of a lack of national guidance, until recently,[6] on effective and consistent planning approaches. National guidance is now forthcoming through recommendations of the Land and Water Forum, and implementation of the National Policy Statement for Freshwater Management (Ministry for the Environment, 2014). However, national minimum standards for water quality are criticized as too lenient, overlooking the requirement that councils must set limits higher than those minima.

Regulatory difficulties are arising in managing diffuse pollution, especially from agriculture and urban runoff. This is essentially a land-management problem due to a mosaic of land uses and requires land-use policy, not just water policy. Some catchments with high levels of irrigation and urban water use in relation to naturally available water have become 'engineered flow systems' – an exemplar of the Anthropocene era (Savenije, Hoekstra, & Van Der Zaag, 2014) – with a need for clearer legal connection between catchment management and water infrastructure planning. This reduced the scores for policy coherence (Principle 3) and water financing (Principle 6) because of differing legal and financing mandates.

Brazilian national water policy

The principles of the Brazilian Water Act and the framework for the National Water Resources Management System are well aligned with the OECD principles. The degree of alignment scores high (5 or 4) for all the principles except Principle 12.

Twenty years since the Water Act was approved there are still gaps regarding implementation and on-ground results. For these two criteria, lower scores are found. As the country is so vast and diverse, the current status of implementation and on-ground results varies across the water planning scales.

Regarding policy impact, water reform at state and river-basin levels was induced, with a water law approved in each state, which is why it scores high (4) for some of the principles. However, in general, lower scores are achieved for policy impact, most being related to trust and engagement. Principle 12 (monitoring and evaluation) has the lowest scores for all four criteria. The National Water Agency publishes annual reports on the Brazilian water resource situation (e.g., ANA, 2016), and there are studies on Brazilian water governance (e.g., OECD, 2015b). However, independent and regular monitoring and evaluation have not been established.

South African national water policy

South Africa's water legislation is highly innovative, and very well aligned with the OECD principles. However, its implementation is problematic, with too many moving goalposts, especially with regard to delegations from national to catchment level. A major policy review and other activities reflect a rollback from decentralization to recentralization. In the past few years there have been several ad hoc, highly politicized, but relatively ineffective initiatives. Some aspects are approached in a participatory manner, but often only with empowered and influential stakeholders.

While lack of clarity on coordination of policies in South Africa remains, particularly around responsibilities for water and environment, many other areas are now well coordinated, 20 years after the initial reforms. The South African law is very clear on sharing and open access to data and information, but many monitoring systems are deteriorating, and data quality control seems to have declined. Enforcement is also difficult to assess. Citizen science is emerging strongly in water governance in many areas of the country, but it is not being recognized as a valid database by government and decision makers.

Managing trade-offs is left to catchment management agencies to negotiate, but only in their advisory capacity. Otherwise, the licensing process happens at the national level, much removed from the area, stakeholders and impacts. Owing to the continually moving goalposts and lack of decision making, there appears to be another era of stakeholder fatigue. There is little assurance or confidence in the government's ability to rectify the inequalities of the past as well to ensure sustainable water management, because the resource and the environment are continuously deteriorating in all parts of the country.

Overall, South Africa's struggle with implementation and on-ground results seems to be mainly due to political issues as well as a variety of capacity challenges in management, decision making and stakeholder engagement.

Lisbon charter

The Lisbon Charter is focused on water services. The drive to innovate in public policies comes from expectations on governments to respond to new frameworks at the international level, including the Sustainable Development Goals and the Human Right to Safe Drinking Water and Sanitation.

Based on application in Portugal, there is a strong alignment and complementarity between the OECD principles, which is more focused on water resources but also includes water services, and the Lisbon Charter, which is more focused on water services but does not disregard water resources. While the OECD principles and the

Lisbon Charter were developed in parallel and approved in 2015 by two different organizations, there were fruitful interactions between them.

The new public policy developed in Portugal for drinking water supply, sanitation and wastewater management services complies with the Lisbon Charter principles and provides a good example of implementation. In the charter's frame, this public policy presents on average a high degree of alignment with the OECD principles, with some variation between principles. Alignment is good regarding Principles 1, 4, 6, 9, 10 and 11, and extremely good regarding Principles 5, 7, 8 and 12. However, it is still insufficient regarding Principle 2, on management scales, and 3, on policy coherence. It is important to note that this analysis is made from the viewpoint of water services and implementation of the Lisbon Charter principles, and not from the viewpoint of water resources management and planning policies, for which results would be different.

Discussion of the 12 OECD principles

This section summarizes the results of the analysis done for each OECD principle across the six frameworks, based on the scoring presented in Appendix 3.

Principle 1: clear roles and responsibilities

There is strong or full alignment between OECD principles and all six frameworks with respect to the need to clearly allocate and distinguish roles and responsibilities. Implementation varies more widely, from poor to full. On-ground results range from poor to good, and the impact on policy was assessed as moderate or strong for all frameworks.

Principle 2: appropriate scales within basin systems

The OECD principles and the six frameworks are in full alignment regarding the need to manage water at appropriate scales and to foster coordination between those scales, with moderate to full implementation. On-ground results are considered moderate or strong in all cases, and policy impact was evenly split between moderate and strong.

Principle 3: policy coherence

There is strong or full alignment for five of the six frameworks on the encouragement of policy coherence through effective cross-sectoral coordination; the other (WFD) was considered moderately aligned. Implementation was assessed as moderate or good for all the frameworks. On-ground results and policy impact both ranged from poor to strong.

Principle 4: capacity

The need to adapt the level of capacity to the complexity of the challenges faced was assessed as strong or full alignment in five of the six frameworks, but the other (South Africa) was considered poor. Implementation is more variable, ranging from poor to full. On-ground results were found to be generally moderate to major changes evident, although for one framework there is no evidence of change. Policy impact varied from none to very strong impact.

Principle 5: produce and share data and information

Five of the six frameworks were found to have full alignment with the OECD principle of producing, updating and sharing consistent, comparable and policy-relevant water and water-related data and information, with the other considered to have strong alignment. Implementation was evenly spread between moderate, strong and full. On-ground results were poor for one framework (South Africa), but ranged from moderate to major changes evident for the others. Policy impact ranged from poor (South Africa) to very strong (Lisbon Charter).

Principle 6: financing

There was moderate to full alignment with the OECD principles of water finance mobilization, and allocating financial resources in an efficient, transparent and timely manner, for all frameworks. Implementation was assessed as moderate or good for five frameworks. On-ground results were generally poor to moderate, with one framework assessed as strong. Policy impact was evenly spread between poor, moderate and strong.

Principle 7: regulatory frameworks

Full alignment with the principle of ensuring sound regulatory frameworks was identified for five of the six frameworks, with another considered in strong alignment. Implementation was evenly spread, with two each at moderate, strong and full implementation. The results for on-ground results were varied, with four of the six frameworks assessed as moderate or good, one (South Africa) as having no evidence of change and the other (Lisbon Charter) as major changes evident. Policy impact ranged from poor to very strong.

Principle 8: adopt and implement innovative governance

Alignment with the OECD principle of promoting innovative water governance practices was good to full for five of the six frameworks, with the other assessed as moderate. Implementation was moderate or strong for most frameworks, with one (Lisbon Charter) assessed as fully implemented for Portugal and one (South Africa) as no implementation. Similarly, on-ground results were considered moderate or strong for most frameworks, with one (Lisbon Charter) assessed as major changes evident and the other (South Africa) as no evidence of change. Policy impact was identified as moderate in four cases and strong in another (Lisbon Charter), but one (South Africa) was assessed as having no impact on policy.

Principle 9: integrity and transparency

There was either strong or full alignment with the OECD principle of mainstreaming integrity and transparency to improve accountability and trust for all except one framework (South Africa), which was considered moderate. Implementation was much more varied, with three considered poor, and the other three either moderate, good or in one case (New Zealand) full. On-ground results were evenly spread between poor, moderate and strong, and policy impact also ranged from poor to strong.

Principle 10: stakeholder engagement

All the policies were found to be in full alignment with the OECD principle of promoting stakeholder engagement, although there were differences in implementation, which was spread between moderate, good and full. On-ground results were identified as moderate to strong, while there were major changes evident in one case (Australia). Similarly, policy impact was generally moderate or good, with one framework (New Zealand) assessed as having very strong impact.

Principle 11: managing trade-offs

Alignment with the OECD principle of encouraging frameworks to help manage intersectoral, spatial and temporal trade-offs was moderate to full for five of the six frameworks, but for one (WFD), it was considered poor. Similarly, Implementation was assessed as moderate or good for five of the frameworks, with the other (WFD) considered poor. On-ground results and policy impacts were both found spread between poor, moderate and strong.

Principle 12: monitoring and evaluation

For five of the six frameworks there was strong to full alignment with the OECD principle of promoting regular monitoring and evaluation of policy and governance, while for the other (Brazil) it was assessed as moderate. There was moderate to full implementation in all but one framework (Brazil), where implementation was considered poor. Similarly, on-ground results were assessed as moderate to major changes evident for five cases, although for one framework (Brazil) it was considered poor. Policy impact was varied, with three frameworks assessed as strong, and one as very strong, but in two cases, poor.

Table 2 presents the highest- and lowest-scoring principles against each of the four assessment criteria. Typically the three highest and three lowest scoring principles are shown; in case of a draw, more are included.

Table 2. Highest- and lowest-scoring principles against each of the criteria.

Criterion	Highest-scoring principles	Lowest-scoring principles
Alignment	2 – Appropriate Scales with Basin Systems 10 – Stakeholder Engagement 5 – Produce and Share Data and Information 7 – Regulatory Frameworks	4 – Capacity 11 – Managing Trade-offs 3 – Policy Coherence 6 – Financing
Implementation	7 – Regulatory Frameworks 10 – Stakeholder Engagement 5 – Produce and Share Data and Information	3 – Policy Coherence 4 – Capacity 9 – Integrity and Transparency 6 – Financing 8 – Adopt and Implement Innovative Governance
On-ground results	10 – Stakeholder Engagement 5 – Produce and Share Data and Information 12 – Monitoring and Engagement	3 – Policy Coherence 6 – Financing 9 – Integrity and Transparency 11 – Managing Trade-offs
Policy impact	10 – Stakeholder Engagement 5 – Produce and Share Data and Information 7 – Regulatory Frameworks	3 – Policy Coherence 11 – Managing Trade-offs 8 – Adopt and Implement Innovative Governance

The OECD principles relating primarily to enhancing the 'effectiveness of water governance' only scored highly twice and were among the lowest scores on six occasions. Notably, policy coherence was one of the lowest-scoring principles against all four assessment criteria. OECD principles relating primarily to enhancing the 'efficiency of water governance' scored highly seven times but were also among the lowest scores on five occasions. 'Produce and share data and information' was one of the highest-scoring principles against all four assessment criteria. OECD principles relating primarily to enhancing 'trust and engagement' in water governance scored highly five times but were also among the lowest scores on five occasions. Stakeholder engagement was one of the highest-scoring principles against all four assessment criteria.

Looking across all the frameworks it is clear that alignment always scored higher than implementation, on-ground results and policy impact. Notably, we found that appropriate scales within basin systems (Principle 2) and stakeholder engagement (Principle 10) were both in full alignment with all the policy frameworks we considered. On the other hand, capacity (Principle 4), managing trade-offs (Principle 11), policy coherence (Principle 3) and financing (Principle 6) were the least aligned with the frameworks we examined.

No principles where considered to be at full implementation for all the frameworks. Implementation of regulatory frameworks (Principle 7), stakeholder engagement (Principle 10) and produce and share data and information (Principle 5) were the best implemented, while integrity and transparency (Principle 9) and policy coherence (Principle 3) were the least. There were no principles for which there were major changes evident for all frameworks. Major changes were most evident in relation to stakeholder engagement (Principle 10), produce and share data and information (Principle 5) and monitoring and evaluation (Principle 12). The least change was found to have occurred in relation to financing (Principle 6), integrity and transparency (Principle 9), managing trade-offs (Principle 11) and regulatory frameworks (Principle 7).

There were no principles for which we considered that there were major impacts on policy for all frameworks. Across the frameworks, the greatest policy impact was in relation to stakeholder engagement (Principle 10), produce and share data and information (Principle 5), regulatory frameworks (Principle 7) and clear roles and responsibilities (Principle 1), whereas policy coherence (Principle 3), managing trade-offs (Principle 11) and adopt and implement innovative governance (Principle 8) seem to have had the least impact.

Conclusions and recommendations

Concluding notes from the analysis

In relation to the OECD principles, this study identified four main target areas for enhancing effectiveness, efficiency, and trust and engagement in water governance. In order of importance, these were policy coherence (Principle 3); financing (Principle 6); managing trade-offs across users, rural and urban areas, and generations (Principle 11); and integrity and transparency (Principle 9). The results of a qualitative analysis undertaken across the findings for each framework, to identify factors needing future action in these four target areas, are presented below.

Policy coherence as a key target for enhancing the effectiveness of water governance

OECD Principle 3 asks us to 'encourage policy coherence through effective cross-sectoral co-ordination, especially between policies for water and the environment, health, energy, agriculture, industry, spatial planning and land use' and identifies four ways to address this.[7] This study found that policy coherence was the only OECD principle that was among the lowest-scoring for all four criteria. It was also the lowest-scoring principle overall.

The qualitative analysis revealed that factors contributing to lower levels of achievement in relation to Principle 3 include: a lack of national guidance on effective and consistent planning approaches; absence of land-use policy; lack of coordination of policy and implementation between water and other policy areas such as the environment; misalignment of policy targets; differing mandates of national, state/regional and local government agencies, and infrastructure and sectoral entities; lack of integration between water quality and quantity in decision making; disconnected approaches between management of services and resources; impact of moving goalposts on implementation; a shift back from decentralization to recentralization; insufficient use of socio-economic data in water-sector planning; overlaps and gaps in implementation despite clear roles and responsibilities; lack of integration of national and state information systems; difficulties implementing complex catchment plans which involve managing use, discharge and cumulative land use within limits; and challenges in coordinating across different scales in large and diverse countries.

While multilevel governance is a clear goal, it raises many challenges in communication, power-sharing and coherence. There is a need for innovative governance methods that better align policy objectives with values consistently across levels and scales.

Financing as a key target for enhancing the efficiency of water governance.

OECD Principle 6 ask us to 'ensure that governance arrangements help mobilize water finance and allocate financial resources in an efficient, transparent and timely manner' and identifies five ways to address this.[8] However, this study found that financing was among the lowest-scoring principles for alignment, implementation and on-ground results, and the fourth-lowest-scoring principle overall.

The qualitative analysis revealed that factors contributing to lower levels of achievement in relation to Principle 6 include: awareness of problems with water finance but lack of clarity on how it will be addressed; lack of transparency in rationales for water pricing; reductions in funding, compromising implementation of innovative practices already identified; implementation capacity restricted by setting fees too low; lack of finance for restoring and rehabilitating polluted water resources; prioritization of spending on infrastructure above other actions such as water demand management; and inaccessible data, declining data quality control and deteriorating monitoring systems.

Insufficient or ineffective financing of water management affects capacity and capability to implement reforms despite there being good legal and policy frameworks in place. To help address this, financing should be focused on more comprehensive water management which integrates water resource and infrastructure planning.

Managing trade-offs across users, rural and urban areas, and generations as a key target for enhancing trust and engagement in water governance.

OECD Principle 11 asks us to 'encourage water governance frameworks that help manage trade-offs across water users, rural and urban areas, and generations' and identifies four ways to address this.[9] However, this study found that managing trade-offs was among the lowest-scoring principles for alignment, on-ground results, and policy impact, and was the third-lowest-scoring principle overall.

The qualitative analysis revealed that factors contributing to lower levels of achievement in relation to this principle include: under-emphasis of intergenerational equity; lack of specifically coordinated action with other sector policies; lack of integration of water management with mining and energy production as strategic water uses; insufficient socio-economic data used for water-sector planning; decisions made in isolation by governments, or with river basin committees restricted to a secondary or advisory role, despite commitments to decentralize decision making; differences in institutional capacity across the country and across water management levels; limited government support to address capacity imbalances; lack of cross-sectoral coordination because different sectors are under different ministries; lack of consistent and comparable data and information to create a national picture of the state of water resources; and decisions being driven by today's priorities, particularly economic drivers, despite frameworks acknowledging future generations.

Clearly, questions of trade-offs raise greater difficulties for basins under significant water stress, including those suffering from increasing rural–urban competition for water. Earlier investment in capacity building towards preparedness and adaptation skills can help address this objective.

Integrity and transparency as a key target for enhancingtrust and engagement in water governance.

OECD Principle 9 asks us to 'mainstream integrity and transparency practices across water policies, water institutions and water governance frameworks for greater accountability and trust in decision-making' and identifies five ways to address this.[10] However, although stakeholder engagement appeared in the highest-scoring group for all criteria and was the highest-scoring principle overall, a key requisite for trust and engagement – integrity and transparency – was one of the lowest-scoring principles for both implementation and on-ground results, and the second-lowest-scoring principle overall.

The qualitative analysis revealed that factors contributing to lower levels of achievement in relation to this principle include: there is much yet to be done to build trust, including in relation to trade-offs that are not always transparent; reductions in 'red tape' done in the name of efficiency have undermined trust in government processes; some national standards (e.g., for water quality) are leading to perceptions that the lowest-common-denominator standard applies everywhere; moving goalposts and reduced commitments to transparency in decision making (especially ad hoc removal of responsibilities and functions of catchment management agencies), leading to stakeholder fatigue and less confidence in government; inconsistent commitment within jurisdictions to delegate decision making and the emergence of a power play between stakeholders and water experts; and limited policy evaluation and limited mechanisms for feedback into law on a systematic or organizational level.

Increasing government commitment to longer-term goals, consistent formulation of long-lasting policies independent of political cycles, and communicating clearly to citizens the aims and measures to be taken, then delivering on them, can create more transparency and ensure greater integrity, leading to both better understanding of real problems and potential solutions, and to mutual commitment from governments and society.

Final recommendations and ways forward

From the assessment of each framework and the detailed analysis that produced the concluding remarks in the previous section, several areas are recommended for consideration in relation to the OECD Principles on Water Governance, organized around the three drivers. Far from any intention of exhausting this discussion, these suggestions aim to open ways for further exploration and debate.

Recommendations for improving effectiveness in water governance

Reinforcing a comprehensive approach to water management. A comprehensive and holistic approach needs to be reinforced, adopting more catchment-based approaches and integrating different levels of jurisdiction. An obvious area that deserves to be elevated in consideration by the relevant authorities is water and land management. Generally, concurrent consideration of some of the OECD principles (e.g., 2, 3, 5, 9 and 11) can help develop a clearer and broader understanding of the problems and the solutions.

Strengthening transboundary cooperation. The OECD principles appear to be primarily focused within national boundaries; but of course many problems appear across boundaries, especially where water bodies span national or state jurisdictions. Although Principles 2 and 5 encourage riparian cooperation and the sharing of information between jurisdictions, the OECD principles lack a mechanism to fully address the need for strengthened transboundary cooperation. There may, for example, be opportunities to link the principles with the Convention on the Law of the Non-navigational Uses of International Watercourses and the Convention on the Protection and Use of Transboundary Watercourses and International Lakes to achieve stronger collaborative governance.[11] The case of the WFD linking water policies across Europe is illustrative. It has had an impact on international cooperation on water governance, revolutionized water law and policy in European countries, and triggered a general move towards better national principles and practices.

Recommendations for improving the efficiency of water governance

Filling the gap for weak national water policies. The principles are more easily accepted if there is a governmental framework that incorporates them, but many nations lack efficient and effective water policies. What possibilities exist for supporting implementation of the OECD principles in such circumstances? These are often the most critical cases, and they will eventually not only impact their own national issues within their border, but also 'export' problems across borders through unregulated services, unfair competition, transboundary water sharing and environmental impacts, and in the market place. Building a 'community of (good) practice' would link countries that share similar governance philosophies, and also allow others that do not yet have robust national water policies to benefit from the OECD principles.

Funding the whole water cycle. This study found that in many contexts there is awareness of problems with water finance but a lack of clarity on how they will be addressed. There is a general tendency to continue focusing on financing expansion of infrastructure above other actions such as water demand management. Inefficient financing of water management affects implementation of reforms, constrains innovation and hampers attempts to build the capacity needed for coming challenges. Concurrent consideration of all aspects of water management is advocated to better address the financial needs of the different dimensions and the system as a whole.

Recommendations for improving trust and engagement in water governance

Networking around good practices. The current water problems in the world are not limited to nations, near borders or even regions – many have become global problems. A global paradigm shift and effective application of the OECD principles demands active networking to share good practices beyond borders and regions – beyond transnationally – in other words, across the whole world. A reference example that also inspired this analysis is the UNESCO-IHP Hydrology for the Environment, Life and Policy programme, which developed a method to share good practices and influence national and regional water policies.

National integrity and guidance with locally devolved decisions. More transparency in policy making may promote more accountability around water management. Considering Principle 11 (managing trade-offs) for example, we can conclude that this is a paradigmatic case of the need to engage all the levels in the problem (from local communities to supra-national). This principle is a good example of how the complexity of water problems demands holistic approaches and actions, including considerations of territorial, temporal and intergeneration continuity. For example, trade-offs between rural and urban needs are one of the major challenges the world faces, and water management is at the core of the problem and the solution.

Notes

1. See http://www.catchmentbasedapproach.org/ for further background on the catchment-based approach.
2. The full agreement can be found at https://www.pc.gov.au/inquiries/current/water-reform/national-water-initiative-agreement-2004.pdf.
3. New Zealand's Resource Management Act was seen as a model when the South African Water Act of 1998, discussed later, was being developed.
4. Further information on SINGREH can be found in Silva, Da, Galvão, Ribeiro, and Andrade (2017), ANA (2016), Barbosa, Mushtaq, and Alam (2016), Sousa Júnior et al. (2016), OECD (2015b), Veiga and Magrini (2013), and Ribeiro, Vieira, and Ribeiro (2012).
5. The Lisbon Charter is available at http://www.iwa-network.org/downloads/1428787191-Lisbon_Regulators_Charter.pdf.
6. See e.g. http://www.mfe.govt.nz/fresh-water/freshwater-management-reforms/clean-water-package-2017.
7. To encourage policy coherence through effective cross-sectoral coordination, especially between policies for water and the environment, health, energy, agriculture, industry, spatial planning and land use, the OECD suggests: (1) encouraging coordination mechanisms within government; (2) fostering coordinated management of use, protection and clean-up of resources; (3) identifying, assessing and addressing barriers; and (4) providing incentives and regulations to mitigate conflicts among sectoral strategies. See OECD (2015a) for full details.
8. To ensure that governance arrangements help mobilize water finance and allocate financial resources in an efficient, transparent and timely manner, the OECD suggests: (1) promoting governance arrangements that help raise revenue for the necessary functions; (2) sector reviews and strategic financial planning to help ensure financing; (3) transparent practices for budgeting and accounting; (4) mechanisms for efficient and transparent allocation of funds; and (5) minimizing unnecessary administrative burdens. See OECD (2015a) for full details.
9. To encourage water governance frameworks that help manage trade-offs across water users, rural and urban areas, and generations, the OECD suggests: (1) promoting non-discriminatory participation in decision making; (2) empowering local authorities and users to identify and address barriers; (3) promoting public debate on the risks and costs; and (4) encouraging evidence-based assessment of the distributional consequences of water-related policies. See OECD (2015a) for full details.
10. To mainstream integrity and transparency practices across water policies, water institutions and water governance frameworks for greater accountability and trust in decision-making, the OECD suggests: (1) promoting legal and institutional frameworks that hold decision makers to account; (2) encouraging norms, codes of conduct or charters on integrity and transparency; (3) establishing clear accountability and control mechanisms; (4) diagnosing and mapping existing or potential drivers of corruption; and (5) adopting multi-stakeholder approaches, tools and action plans to address water integrity and transparency gaps. See OECD (2015a) for full details.
11. Convention on the Law of the Non-navigational Uses of International Watercourses 1997, adopted by the General Assembly of the United Nations, 21 May 1997 (http://legal.un.org/ilc/texts/instruments/english/conventions/8_3_1997.pdf); Convention on the Protection and Use of Transboundary Watercourses and International Lakes, Helsinki, 17 March 1992 (https://www.unece.org/fileadmin/DAM/env/water/pdf/watercon.pdf).

Disclosure statement

No potential conflict of interest was reported by the authors.

Funding

M. Ribeiro would like to thank the support from CAPES Foundation – Brazil [Grant no. 426/2016-04].

ORCID

Susana Neto http://orcid.org/0000-0001-5231-8633

References

Akmouch, A., & Correia, F. N. (2016, December). The 12 OECD principles on water governance – When science meets policy. *Utilities Policy*, *43*(Part A), 14–20. doi:10.1016/j.jup.2016.06.004

ANA. (2016). *Conjuntura dos recursos hídricos no Brasil – Informe 2016* [Framework of water resources in Brazil – note 2016]. Brasília: Agência Nacional de Águas. [online] Retrieved from http://www3.snirh.gov.br/portal/snirh/centrais-de-conteudos/conjuntura-dos-recursos-hidricos/informe-conjuntura-2016.pdf

Barbosa, M. C., Mushtaq, S., & Alam, K. (2016). Rationalising water policy and the institutional and water governance arrangements in Sao Paulo, Brazil. *Water Policy*, *18*(6), 1353–1366. doi:10.2166/wp.2016.233

Biswas, A. K., & Tortajada, C. (2010). Future water governance: Problems and perspectives. *International Journal of Water Resources Development*, *26*(2), 129–139. doi:10.1080/07900627.2010.488853

Blackstock, K. L., Martin-Ortega, J., & Spray, C. J. (2015). Implementation of the European water framework directive: What does taking an ecosystem services-based approach add? In J. Martin-Ortega, R. C. Ferrier, I. J. Gordon, & S. Khan (Eds.), *Water ecosystem services: A global perspective*. International Hydrology Series. (pp. 57–64). Cambridge: Cambridge University Press.

Chenoweth, J. L. (1998). Conflict in water use in Victoria, Australia: Bolte's divide, *Australian Geographical Studies*, *36*, 248–253.

Connell, D., Robins, L., & Dovers, S. (2007). Delivering the national water initiative: Institutional roles, responsibilities and capacities. In K. Hussey & S. Dovers (Eds.), *Managing water For Australia: The social and institutional challenges*. Collingwood: CSIRO Publishing.

Correia, F. N. (2003, October 5–9). *Institutional water issues in Europe*. Invited keynote paper to the XI World Water Congress of IWRA, Madrid Congress Hall, Madrid, Spain (pp. 595–596). doi:10.1080/02508060208687049

De la Harpe, J., & Ramsden, P. (1998). *Guide to the National Water Act*. Pretoria: Department of Water Affairs and Forestry.

Duncan, R. (2014). Regulating agricultural land use to manage water quality: The challenges for science and policy in enforcing limits on non-point source pollution in New Zealand. *Land Use Policy*, *41*, 378–387. doi:10.1016/j.landusepol.2014.06.003

European Commission. (1993). *Proposal for a Council Directive on the ecological quality of water. COM (93) 680 final. 15 June 1994* [EU Commission - COM Document].

European Commission. (2017). *Status of implementation of the WFD in the Member States*. European Commission. Last updated 28/10/2016. Retrieved February 23, 2017, from http://ec.europa.eu/environment/water/participation/map_mc/map.htm

Evans, M., & Pratchett, L. (2013). The localism gap – The CLEAR failings of official consultation in the Murray Darling basin. *Policy Studies*, *34*, 541–558. doi:10.1080/01442872.2013.862448

Fenemor, A., Neilan, D., Allen, W., & Russell, S. (2011). Improving water governance in New Zealand stakeholder views of catchment management processes and plans. *Policy Quarterly*, *7*(4), 10–19.

Fenemor, A. D. (2017). Water governance in New Zealand - challenges and future directions. *New Water Policy and Practice*, *3*(1), 9–21. doi:10.18278/nwpp.3.1.3.2.2

Fenemor, A. D., Davie, T., & Markham, S. (2006). Hydrological information in water law and policy: New Zealand's devolved approach to water management. Chapter 12. In J. Wallace & P. Wouters (Eds.), *Hydrology and water law - Bridging the gap*. London: IWA Publishing.

Gentle, G., & Olszak, C. (2007). Water planning: Principles, practices and evaluation. In K. Hussey & S. Dovers (Eds.), *Managing water for Australia: The social and institutional challenges*. Collingwood: CSIRO Publishing.

Harmsworth, G. R., Young, R. G., Walker, D., Clapcott, J. E., & James, T. (2011). Linkages between cultural and scientific indicators of river and stream health. *New Zealand Journal of Marine and Freshwater Research, 45*(3), 423–436. doi:10.1080/00288330.2011.570767

Hendry, S. (2014). Frameworks for water law reform. In *Frameworks for water law reform* (International Hydrology Series, p. I). Cambridge: Cambridge University Press. ISBN 978-1-107-01230-1.

Hering, D., Borja, A., Carstensen, J., Carvalho, L., Elliott, M., Feld, C. K., … van de Bund, W. (2010, September). The European water framework directive at the age of 10: A critical review of the achievements with recommendations for the future. *Science of the Total Environment Science of the Total Environment, 408*(19), 4007–4019. doi:10.1016/j.scitotenv.2010.05.031

Hussey, K., & Dovers, S. (Eds.). (2007). *Managing water for Australia: The social and institutional challenges*. Collingwood: CSIRO Publishing.

IWA. (2015). *Lisbon charter for guiding the public policy and regulation of drinking water supply*. London: Sanitation and Wastewater Management Services.

Jackson, T., & Dixon, J. (2007). The New Zealand resource management Act: An exercise in delivering sustainable development through an ecological modernisation agenda. *Environment and Planning B: Planning and Design, 34*(1), 107–120. doi:10.1068/b32089

Jager, N. W., Challies, E., Kochskämper, E., Newig, J., Benson, D., Blackstock, K., … Von Korff, Y. (2016). Transforming European water governance? Participation and River Basin management under the EU water framework directive in 13 member states. *Water, 8*, 156, 1–23. doi:10.3390/w8040156

Kildea, P., & Williams, G. (2010). The constitution and the management of water in Australia's rivers. *Sydney Law Review, 32*, 595.

Land and Water Forum. (2012). *Second report of the land and water forum: setting limits for water quality and quantity, and freshwater policy- and plan-making through collaboration*. [online] Retrieved from http://www.landandwater.org.nz/

Lloyd, C. J. (1988). *Either drought or plenty: Water development and management in New South Wales*. Parramatta: Department of Water Resources NSW.

Meijerink, S., & Huitema, D. (2010). Policy entrepreneurs and change strategies: Lessons from sixteen case studies of water transitions around the globe. *Ecology And Society 15*(2), 21. Retrieved from http://www.ecologyandsociety.org/vol15/iss2/art21/

Ministry for the Environment. (2014). *A guide to the national policy statement for freshwater management 2014*. [online] Retrieved from http://www.mfe.govt.nz/publications/fresh-water/guide-national-policy-statement-freshwater-management-2014

Monaghan, R. M., De Klein, C., & Muirhead, R. W. (2008). Prioritisation of farm scale remediation efforts for reducing losses of nutrients and faecal indicator organisms to waterways: A case study of New Zealand dairy farming. *Journal of Environmental Management, 87*(4), 609–622. doi:10.1016/j.jenvman.2006.07.017

National Water Commission. (2004). *Intergovernmental agreement on a national water initiative* (39pp). Canberra: Author.

National Water Commission. (2010). *Australian environmental water management report 2010*. Canberra: NWC.

National Water Commission. (2014a). *Fourth biennial assessment of the national water initiative*. Canberra: Author.

National Water Commission. (2014b). *National water planning report card 2013*. Canberra: Author.

Neto, S. (2010). Water, territory and planning. Contemporary challenges: towards a territorial integration of water management (PhD Thesis). [online] Retrieved from https://www.researchgate.net/profile/Susana_Neto2/publication/282850166_Water_Territory_and_Planning_

Contemporary_Challenges_towards_a_Territorial_Integration_of_Water_(PhD_Abstract)/links/561e642708aef097132c4243.pdf

OECD. (2015a). OECD principles on water governance welcomed by ministers at the OECD ministerial council meeting on 4 June 2015. Directorate for Public Governance and Territorial Development. Paris: Author. Retrieved from http://www.oecd.org/gov/regional-policy/OECDPrinciples-on-Water-Governance-brochure.pdf.

OECD. (2015b). *Water resources governance in Brazil.* Paris: Author. doi:10.1787/9789264238121-en

Pahl-Wostl, C. (2009). A conceptual framework for analysing adaptive capacity and multi-level learning processes in resource governance regimes. *Global Environmental Change, 19*, 354–365. doi:10.1016/j.gloenvcha.2009.06.001

Pigram, J. J. (2006). *Australia's water resources: From use to management.* Vic: CSIRO Publishing.

Powell, J. M. (1989). *Watering the garden state: Water, land and community in Victoria 1834-1988.* North Sydney: Allen Uniwin.

Ribeiro, M. F., Vieira, M. C., & Ribeiro, M. R. (2012). Participatory and decentralized water resources management: Challenges and perspectives for the North Paraiba River Basin Committee-Brazil. *Water Science and Technology, 66*(9), 2007–2013. doi:10.2166/wst.2012.414

Rouillard, J., & Spray, C. (2016, Jun 16). Working across scales in integrated catchment management: Lessons learned for adaptive water governance from regional experiences. *Regional Environmental Change.* doi:10.1007/s10113-016-0988-1)

Rouse, H. L., & Norton, N. (2016). Challenges for freshwater science in policy development: Reflections from the science-policy interface in New Zealand. *New Zealand Journal of Marine and Freshwater Research, 51*, 1–14.

Ruru, J. (2013). Indigenous restitution in settling water claims: The developing cultural and commercial redress opportunities in Aotearoa, New Zealand. *Pacific Rim L & Pol'y Journal, 22*, 311.

Salmond, A. (2014). Tears of rangi: Water, power, and people in New Zealand. *HAU: Journal of Ethnographic Theory, 4*(3), 285–309. doi:10.14318/hau4.3

Savenije, H. G., Hoekstra, A. Y., & Van der Zaag, P. (2014). Evolving water science in the Anthropocene. *Hydrology and Earth System Sciences, 18*(1), 319–332. doi:10.5194/hess-18-319-2014

Silva, C. S., Da, Galvão, C. O., Ribeiro, M. R., & Andrade, T. S. (2017). Adaptation to climate change: Institutional analysis. Chapter 10. In E. Lolokytha, S. Oishi, & R. S. V. Teegavarapu (Eds.), *Sustainable water resources planning and management* (pp. 261–279). Singapore: Springer.

Sousa Júnior, W., Baldwin, C., Camkin, J., Fidelman, P., Silva, O., Neto, S., & Smith, T. F. (2016). Water: Drought, crisis and governance in Australia and Brazil. *Water, 8*(11), 493. doi:10.3390/w8110493

Spray, C., & Comins, L. (2011). 'Governance structures for effective integrated catchment management: Lessons from the Tweed HELP Basin. *UK' Journal of Hydrologic Environment, 7*(1), 105–109.

State of Victoria Department of Environment, Land, Water and Planning. (2016). *Water for Victoria water plan.* Melbourne: State of Victoria Department of Environment, Land, Water and Planning.

Tan, P.-L. (2002). *Legal issues relating to water use, issues paper no.1*, Murray-Darling Commission Project MP2002, Report to the Murray-Darling Basin Commission, reprinted in *Property: Rights and Responsibilities, Current Australian Thinking*, Land and Water. Canberra, Australia.

Tan, P.-L., & Auty, K. (2017). Finding diamonds in the dust: Community engagement in Murray- Darling Basin planning. In B. Hart & J. Doolan (Eds.), *Decision making in water resources policy and management.* London: Elsevier.

Tan, P.-L., Bowmer, K., & Mackenzie, J. (2012). Deliberative tools for meeting the challenges of water planning in Australia. *Journal of Hydrology, 474*, 2–10. doi:10.1016/j.jhydrol.2012.02.032

Tan, P.-L., & Jackson, S. (2013). Impossible dreaming - Does Australia's law and policy fulfil indigenous aspirations? *Environmental and Planning Law Journal, 30*, 132–149.

Taylor, A., Wouter, T., Arriëns, L., & Laing, M. (2015). Understanding six water leadership roles: A framework to help build leadership capacity. *New Water Policy and Practice*, *1*, 4–31. doi:10.18278/nwpp.1.2.2

UNESCO. (2010). HELP: Hydrology for the environment, life and policy. Author. Paris, France. Retrieved January 25, 2017, from http://unesdoc.unesco.org/images/0021/002145/214516E.pdf.

Veiga, B. E., & Magrini, A. (2013). The Brazilian water resources management policy: Fifteen years of success and challenges. *Water Resources Management*, *27*(7), 2287–2302. doi:10.1007/s11269-013-0288-1

Woodhouse, P., & Muller, M. (2017). Water governance - an historical perspective on current debates. *World Development*, *92*(1), 225–241. doi:10.1016/j.worlddev.2016.11.014

Appendix 1

The 12 OECD principles of water governance

Enhancing the *effectiveness* of water governance

(1) Clearly allocate and distinguish roles and responsibilities for water policymaking, policy implementation, operational management and regulation, and foster co-ordination across these responsible authorities.
(2) Manage water at the appropriate scale(s) within integrated basin governance systems to reflect local conditions, and foster co-ordination between the different scales.
(3) Encourage policy coherence through effective cross-sectoral co-ordination, especially between policies for water and the environment, health, energy, agriculture, industry, spatial planning and land use.
(4) Adapt the level of capacity of responsible authorities to the complexity of water challenges to be met, and to the set of competencies required to carry out their duties.

Enhancing the *efficiency* of water governance

(1) Produce, update, and share timely, consistent, comparable and policy-relevant water and water-related data and information, and use it to guide, assess and improve water policy.
(2) Ensure that governance arrangements help mobilise water finance and allocate financial resources in an efficient, transparent and timely manner.
(3) Ensure that sound water management regulatory frameworks are effectively implemented and enforced in pursuit of the public interest.
(4) Promote the adoption and implementation of innovative water governance practices across responsible authorities, levels of government and relevant stakeholders.

Enhancing *trust and engagement* in water governance

(1) Mainstream integrity and transparency practices across water policies, water institutions and water governance frameworks for greater accountability and trust in decision-making.
(2) Promote stakeholder engagement for informed and outcome-oriented contributions to water policy design and implementation.
(3) Encourage water governance frameworks that help manage trade-offs across water users, rural and urban areas, and generations.
(4) Promote regular monitoring and evaluation of water policy and governance where appropriate, share the results with the public and make adjustments when needed.

Source: OECD (2015a).

Appendix 2
Adapted Likert scale descriptors for scoring each policy framework

Alignment

(1) No alignment
(2) Poor = some common objectives
(3) Moderate = common objectives and measures of policy proposed
(4) Good/strong = previous experience and well-aligned policy ongoing
(5) Full alignment = policy framework matching all the objectives of the OECD principle

Implementation

(1) No implementation
(2) Poor = minimally addressed
(3) Moderate = consistently included, with some measures proposed
(4) Good/strong = under implementation through measures in place
(5) Full alignment = implemented with evaluated results/good practice

On-ground results

(1) No evidence of change
(2) Poor = involving major agent of change (institutional or other)
(3) Moderate = involving different agencies and stakeholders
(4) Good/strong = involving multilevel platforms of participation and decision making
(5) Major changes evident = implemented with evaluated results/good practice

Policy impact

(1) No impact
(2) Poor = considered and being implemented in the ongoing water policy
(3) Moderate = considered for implementation in other policies (transversal impact)
(4) Good/Strong = impacting different institutional levels of governance (vertical impact, bottom-up and top-down)
(5) Very strong impact = producing political change after evaluation (e.g., new legislation, regulatory measures, institutional restructuring or innovative institutional arrangements)

Appendix 3
Results of scoring for each framework

Frameworks / OECD principles	Criteria	Europe WFD	Australia NWI	New Zealand	Brazil	South Africa	Lisbon Charter (Illustrative case: Portugal)
Principle 1: Clear roles and responsibilities)	Alignment	4	5	5	5	4	5
	Implementation	2	4	5	4	3	4
	On-ground results	2	4	4	3	3	4
	Policy impact	3	4	4	4	3	4
Principle 2: Appropriate scales within basin systems)	Alignment	5	5	5	5	5	5
	Implementation	3	5	5	3	3	3
	On-ground results	3	4	4	3	3	3
	Policy impact	4	4	4	3	3	3
Principle 3: Policy coherence)	Alignment	4	4	4	5	3	5
	Implementation	3	4	3	3	3	4
	On-ground results	3	4	4	3	2	3
	Policy impact	3	3	3	4	2	3
Principle 4: Capacity)	Alignment	4	5	4	4	2	5
	Implementation	3	4	5	2	2	4
	On-ground results	3	4	5	3	1	4
	Policy impact	3	3	4	3	2	5
Principle 5: Produce and share data and information)	Alignment	5	5	5	5	4	5
	Implementation	3	5	4	3	3	5
	On-ground results	3	5	5	3	2	4
	Policy impact	4	4	3	4	2	5
Principle 6: Financing)	Alignment	3	5	4	5	3	5
	Implementation	2	4	4	3	3	4
	On-ground results	2	4	3	3	2	4
	Policy impact	2	4	4	3	2	4
Principle 7: Regulatory frameworks)	Alignment	5	5	5	5	4	5
	Implementation	3	5	5	4	3	4
	On-ground results	3	4	3	3	1	5

(Continued)

(Continued).

Frameworks / OECD principles	Criteria	Europe WFD	Australia NWI	New Zealand	Brazil	South Africa	Lisbon Charter (Illustrative case: Portugal)
	Policy impact	4	4	4	3	2	5
Principle 8: Adopt and implement innovative governance)	Alignment	4	5	5	5	3	5
	Implementation	3	4	4	3	1	5
	On-ground results	3	4	3	3	2	5
	Policy impact	3	3	3	3	1	5
Principle 9: Integrity and transparency)	Alignment	4	5	5	4	3	5
	Implementation	2	3	5	2	2	4
	On-ground results	2	3	4	3	2	4
	Policy impact	3	3	4	2	2	4
Principle 10: Stakeholder engagement)	Alignment	5	5	5	5	5	5
	Implementation	3	4	5	3	4	4
	On-ground results	3	4	4	3	4	4
	Policy impact	3	4	5	3	4	4
Principle 11: Managing trade-offs)	Alignment	2	5	4	5	3	5
	Implementation	2	4	4	3	3	4
	On-ground results	2	4	4	3	2	3
	Policy impact	2	4	4	3	2	3
Principle 12: Monitoring and evaluation)	Alignment	5	5	4	3	4	5
	Implementation	3	5	4	2	3	5
	On-ground results	3	5	3	2	4	5
	Policy impact	4	4	4	2	2	5

Functions of OECD Water Governance Principles in assessing water governance practices: assessing the Dutch Flood Protection Programme

Chris Seijger, Stijn Brouwer, Arwin van Buuren, Herman Kasper Gilissen, Marleen van Rijswick and Michelle Hendriks

ABSTRACT

The OECD Principles on Water Governance aim to contribute to good water governance. Learning and change through assessments are useful ways to strengthen water governance systems. This article presents a methodology for a learning assessment based on the OECD principles. The methodology has been applied to the Dutch Flood Protection Programme. The analysis revealed various functions of the OECD principles, from enhancing understanding to reforming the agenda, reflection and informed action. Recommendations are given on how the OECD principles can be used to come to meaningful action-oriented water governance assessments; they include contextualization, multiple methods, inclusiveness and periodic assessments.

Enhancing the performance of water governance systems

The world is increasingly confronted with water crises. Soaring populations, expanding cities and growing economies not only put the world's water resources under pressure, but also magnify the impact of water-related disasters such as floods, droughts and contaminated water supplies (Asian Development Bank, 2016; United Nations World Water Assessment Programme, 2015). Recent examples are the floods in Southern India (2015, over 500 casualties, damage USD 3–15 billion), persistent droughts affecting Brazil's megacity São Paolo (2015, 20 million inhabitants), and the spread of cholera through contaminated water in West and Central Africa (2014–2015, over 90,000 reported cases, over 1600 casualties). Water disasters also have an impact in Europe. A recent evaluation of floods in Germany (2013, damage €6–8 billion) suggests that there is ample space for improvement in the governance system (Thieken et al., 2016). As water crises are affecting billions of people and their living environment, it is not surprising that water crises are topping the list of global risks with highest concern for the coming 10 years (World Economic Forum, 2016).

Current water governance systems have thus far not been able to prevent these crises, and the challenges they should address will only magnify (Driessen, Hegger, Bakker, van Rijswick, & Kundzewicz, 2016; Pahl-Wostl, 2009; World Water Assessment Programme, 2003). Water crises are therefore also referred to as water governance crises, as improvements in the performance of water governance systems are needed (World Water Assessment Programme, 2003). Much debate on what constitutes good water governance has been taking place in academic circles and the water community (Huitema et al., 2009; Lautze, De Silva, Giordano, & Sanford, 2011). Although these debates have clarified key dimensions of water governance, such as effectiveness, efficiency and legitimacy, they have not yet resulted in a comprehensive framework to evaluate and redesign water governance systems. As guidance, let alone consensus, was lacking on what constitutes good water governance, water governance principles were developed in 2015 by the OECD Water Governance Initiative,[1] an international multi-stakeholder network of members from the public, private and not-for-profit sectors (Akhmouch & Clavreul, 2016; OECD, 2015). These principles are an attempt to contribute to legitimate, effective and efficient water governance systems that can manage 'too much', 'too little' and 'too dirty' water in a sustainable and inclusive way. Such outcomes could be achieved when the principles are used in the design and implementation of robust water policies.

Although Water Governance Initiative working groups are making progress towards developing water governance indicators and collecting best practices, there remains a need to better understand how the principles could be applied in practice to generate recommendations for enhancing the legitimacy, effectiveness, efficiency and inclusiveness of existing water governance systems. A useful way to strengthen water governance systems is a process of reflexive learning and changing (Gupta et al., 2010; Hajer et al., 2015; Pahl-Wostl, 2009; Termeer et al., 2011). Reflexive learning combines learning and real-life action. Reflections on what has been learned are used to come to well-informed actions. It can provide insights about which elements of a governance system perform well and which should be improved (Ison, Röling, & Watson, 2007; van Rijswick, Edelenbos, Hellegers, Kok, & Kuks, 2014). In addition, learning can help move actors out of the entrenched positions that may typify a suboptimal water governance system (Pahl-Wostl et al., 2007). Given the importance of learning for strengthening water governance systems, the aim of this article is to explore the practical value of the OECD Water Governance Principles for assessing water governance practices, and to present a method for applying them as a tool for reflexive learning. The method is illustrated through a case study of the Dutch Flood Protection Programme, a multi-billion-euro programme to prevent the Netherlands from being flooded. The article ends with reflection on how the principles could serve as instrument to enhance the performance of water governance systems.

Conceptual framework

The OECD Principles on Water Governance

There is a plethora of definitions of water governance (for an overview see Havekes et al., 2016; Lautze et al., 2011; Teisman, van Buuren, Edelenbos, & Warner, 2013). The OECD (2015) has defined water governance as 'the range of political, institutional and administrative rules, practices and processes (formal and informal) through which decisions are taken and implemented, stakeholders articulate their interests and have their concerns considered, and decision-makers are held accountable in the management of water resources and the delivery of water services'.

The OECD Water Governance Principles are based on three pillars, derived from the multitude of water governance definitions: effectiveness, in terms of meeting policy goals and targets at different levels; efficiency, to ensure that benefits and welfare of sustainable water management are achieved through the lowest societal costs reasonably possible; and trust and engagement, as inclusiveness of stakeholders enhances public confidence, fairness and equity. As Figure 1 shows, each pillar is represented by four principles.

Although the OECD principles are meant to contribute to tangible and outcome-oriented water governance policies, the different potential practical functions of these principles have not yet been explicitly formulated by the OECD (2015), except for some indications. But by closely following the discussion within the Water Governance Initiative, we can discern at least four such functions.

The first can be seen as a (soft) strategy for policy coordination. The mere fact that the principles have been endorsed by 42 countries and over 140 major stakeholder groups makes them a common frame of reference, which can be used to fuel the debate

Figure 1. Summary of the OECD Principles on Water Governance (OECD, 2015).

over good water governance and to stimulate governments to strengthen their governance in this domain.

Second, within the OECD Water Governance Initiative, the principles are also used to collect and diffuse best practices for good water governance. These best practices highlight situations in which one or various principles are implemented in an exemplary way. They can thus serve as a source of inspiration for other countries to improve their water governance.

Third, the OECD Secretariat may use the principles in future reviews and recommendations for water governance systems in OECD countries, as well as in in-depth national multi-stakeholder policy dialogues such as those carried out for Mexico (2013), the Netherlands (2014), Jordan (2014), Tunisia (2014) and Brazil (2015). In these assessments the principles can be used as a yardstick to assess whether water governance systems fulfil all relevant functions properly. Moreover, they can be used to formulate recommendations to implement the principles further, when practices fall short.

The fourth function of the principles we distinguish builds on the above assessment function, and is to enhance learning and reflection among participants in water governance systems. When participants reflect on their practices and choices, a reflexive dialogue is initiated which could stimulate understanding, learning and change. This function thus far is the least developed and explored, so it is elaborated on in this article.

Functions of governance assessments

The functions of the principles presented above roughly reflect the more general functions of governance assessment frameworks as discussed in the policy and programme evaluation literature (Hill & Hupe, 2002; Royse, Thyer, & Padgett, 2010). In general, governance frameworks can be used in three different ways. First of all, they can have an auditing function. In that case, frameworks are used to check whether governance systems provide several functions and thus meet the criteria of an (independent, external) auditor (Davies, 1999). Audits can have a more technical orientation, e.g. 'ticking boxes' to find out whether specific functions are in place. They can also have a more qualitative logic, in which audits are used to say something more about how these functions perform. Audits can be conducted by peers (internal audits) or by external actors. In both forms, they are often used as a soft governance tool to enhance policy implementation and coordination.

One step further than the auditing function is the evaluative function of governance assessment frameworks, in which the framework is used to investigate the quality of governance systems or to report whether they have made sufficient progress in improving their functioning. This type of evaluation is often used to facilitate formal accountability procedures (Royse et al., 2010).

The third function of governance assessment frameworks is to enable and stimulate learning and reflection. In that case, frameworks are used to get more insight into the functioning of governance systems and to come to a more detailed analysis of their functionalities and dysfunctionalities. This information is not used to conclude about the system as such, but to start a dialogue about the question of how this information can be used as a basis for change and development (Edelenbos & van Buuren, 2005; van Rijswick et al., 2014). However, to become an effective instrument for learning, several requirements have to be fulfilled. These will be elaborated in the next section on the functions and conditions of learning.

Reflexive learning in governance assessments

Learning can take different forms and may take place at the individual or organizational level. We acknowledge that several learning theories exist, such as transformative learning, and single-, double- and triple-loop learning (Argyris & Schön, 1978; Mezirow, 1995; Romme & Van Witteloostuijn, 1999). As explained, this article focuses on functions related to reflexive learning and change. On a very basic level, learning can have an instrumental function to acquire new knowledge or skills, for instance through communication with others (Mezirow, 1995). In addition, learning may lead to changes in attitude, behaviour and norms, as well as enhancing trust, respect and shared goals (Ison et al., 2007; Reed et al., 2010). Furthermore, reflection and collective action may be another function of learning when working together to improve environmental management (Keen, Brown, & Dyball, 2005). Lastly, a function of learning can be to enhance reform and change in the governance or water system when outcomes of learning result in significant changes in institutional and technical contexts (Pahl-Wostl et al., 2007).

From the literature we can distil several conditions for facilitating learning processes. A first obvious one is that processes should be participatory, with multi-actor interactions (Pahl-Wostl et al., 2007). It is also suggested that the learning context should represent relational qualities such as trust, reciprocity and willingness for mutual understanding (Pahl-Wostl et al., 2007). Moreover, the context for learning has to be open and inclusive, with a diverse set of participants. Ideally, the participants should reflect the variety of perspectives and interests in the system (requisite variety) (Jessop, 2003). In fact, Bressers, Bressers, Kuks, and Larrue (2016) suggest that by involving potential critics a wider scope of the assessment can be secured. Furthermore, it is considered beneficial when a clear-cut issue is addressed and informal actor platforms are installed. In addition, the institutional setting should provide opportunities to learn and to change governance practices. Both can be achieved when powerful organizations acknowledge the need or importance for change and learning by monitoring (Sabel, 1994), and the institutional setting is stable, without being rigid and inflexible (Pahl-Wostl et al., 2007).

At the same time, governance assessments have to be authoritative and independent, and fit the (scientific) criteria of validity and reliability, which presupposes a certain degree of independence and distance. An important question is thus how to make an assessment that has both 'substantial' and 'processual' qualities. One way to combine both qualities is to use methodological triangulation: to use more independent methods for data collection (e.g. survey, interviews) as well as more participatory methods (e.g. round tables, focus groups). Another way is to distribute different roles between the experts involved: to distinguish between experts responsible for the investigation, and experts responsible for coaching and facilitating the learning process (Edelenbos & van Buuren, 2005). More generally, it is necessary to organize a continuous iteration between science and practice to enhance the validity and reliability of the results.

Research design

To explore the practical and reflexive value of the principles, we conducted a case study on the water governance of the Dutch Flood Protection Programme. After introducing this programme, this section presents the method of a learning assessment.

The Flood Protection Programme

The Netherlands is vulnerable to floods. More than half of the country is at risk of flooding, and potential consequences are severe, as about 9 million people live and work in this area, representing an estimated total flood damage risk of about 150 billion euros by 2050 (Kind, 2014). To protect the Netherlands from flooding, a sophisticated system of dikes, locks and storm-surge barriers has been constructed over the past centuries. This includes about 3800 km of primary flood defences protecting the land from flooding from the sea, rivers and big lakes (Figure 2). Technical safety standards for these flood defences have been set in national legislation (Water Act 2009, amended 2017). To maintain and reinforce this flood protection system, preventing the Dutch from being flooded, all flood defence structures are periodically assessed against these standards. After the first two assessments (in 2001 and 2006), two Flood Protection Programmes were undertaken. In these programmes the Dutch government was solely financially responsible for the execution of the programme. Reinforcement projects had to be plain, effective and robust (Seijger, Dewulf, Otter, & Van Tatenhove, 2013). Yet the reinforcement projects became increasingly expensive and construction works were delayed. Therefore several changes were made in the governance of the third and most recent Flood Protection Programme, launched in 2014.

This third Flood Protection Programme is the empirical research object of this article. The initial scope of this programme is the reinforcement of 748 km of primary flood defence structures. The programme has a budget of roughly €4 billion for 2014–2028, to be spent in projects across the country (see Figure 2 for projects up to 2022). The programme is an alliance of the regional water authorities and the national Ministry of Infrastructure and Environment. Key tasks of the programme are to implement new flood risk standards, distribute funds to the projects, monitor and report progress to the Ministry and Parliament, initiate applied research, disseminate knowledge, and build capacities of flood control professionals (e.g. risk-based strategies, project management). Although it provides a basis for (partly) financing the necessary flood defence measures, the programme does not take over the legal responsibility of the individual water authorities for meeting the required safety standards. Thus, the water authorities remain individually responsible for the implementation of the reinforcement projects (Jorissen, Kraaij, & Tromp, 2016; van Rijswick & Havekes, 2012).

With the start of the programme, several changes in the governance of flood protection were introduced (Jorissen et al., 2016):

(1) Collaboration between national and regional water authorities was strengthened by forming an alliance. They have a shared responsibility for the

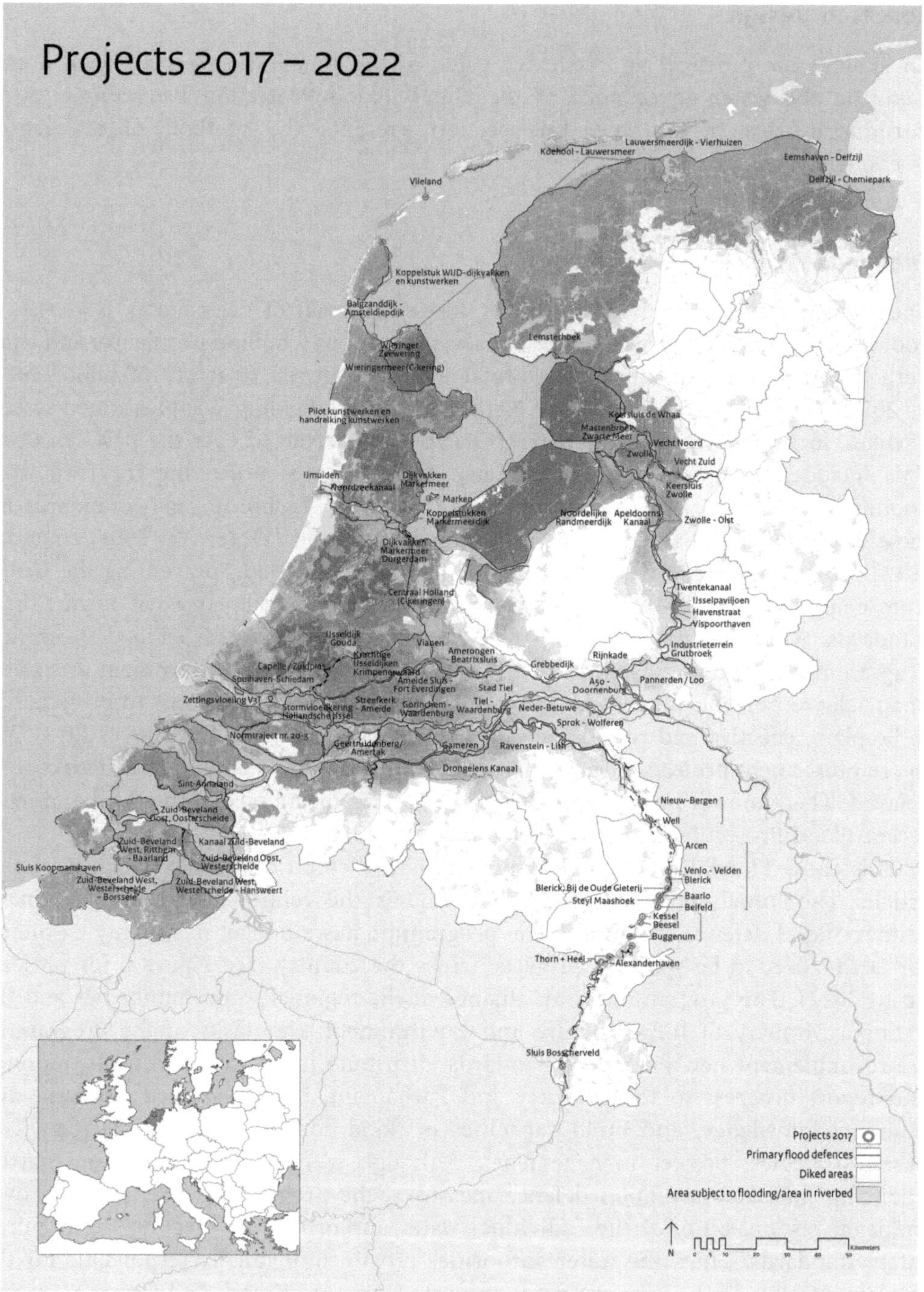

Figure 2. Projects programmed in the Dutch Flood Protection Programme, 2017–2022.

programme. For example, decision making takes place in a joint steering group, and issues that arise in multiple projects are addressed at the programme level.

(2) Costs are shared more equally. In a dike-reinforcement project the national government pays 50%, the collective of regional water authorities 40%, and the responsible regional water authority 10%.
(3) Projects are nationally prioritized in order of urgency (highest risk first). In the previous programmes this was not a criterion for the order of projects.
(4) The programme is no longer static, but is revised annually to actively respond to new insights. Each year a plann for the next six years is made, including a 12-year outlook. There is no fixed deadline or budget for the programme, reflecting that flood protection is a continuous effort.
(5) The programme has set ambitious targets. Compared to the previous Flood Protection Programme(s), the new programme aims to double the output of kilometres of dikes reinforced per year, and to reduce the average cost per kilometre by 30%.

Method

In line with the 'substantial' and 'processual' qualities in a learning assessment (mentioned earlier), our design included independent and more participatory methods for data collection. To maximize the learning and implementation potential of the assessment's findings, in each step experts of the Flood Protection Programme were involved, while at the same time it was ensured that our independence was not compromised. As summarized in Table 1, the design comprises four steps (see the supplemental online material at https://doi.org/10.1080/02508060.2018.1402607 for an overview of participants in the four steps). We recognize that the participants represent a narrow sample. Although this stimulated learning in a safe environment, it may have constrained the representativeness of the assessment outcomes. We further reflect on this limitation in the conclusions.

In Step 1, the scope of the assessment, the main activities, and the criteria for participant involvement were defined. Initially, it was agreed that the assessment would adopt a multilevel focus, including both the national Flood Protection Programme as a whole, plus one selected local dike-reinforcement project (as part of the programme). However, due to lack of response from the selected dike-reinforcement project in the initial steps of the assessment, we decided to exclusively assess the governance of the Flood Protection Programme.

In Step 2, a survey was distributed asking how the Flood Protection Programme performs in the light of the principles. We pre-tested a draft survey with six water governance experts, and their input was used in revising the survey into its final form.

Table 1. The learning assessment consists of four steps.

1. Problem definition	2. Assessment	3. External validation	4. Learning
• Defining objectives, focus, key activities and respondents • Desk study • Coordination with programme	• Assessment of programme according to the OECD Water Governance Principles • Assessment of the programme's results (effectiveness, efficiency, legitimacy) • Online survey • Focus group	• External validation of the results of the assessment • Desk study • Expert interviews	• Systematic inventory of the lessons that can be drawn from the assessment • Learning table with scientists and experts

In the survey, each principle was translated and operationalized for the case-study context, followed by the question of to what extent the principle was relevant to the programme and how the programme performed in the light of that principle. The supplemental online material contains a small sample of (translated) survey questions. The survey was exclusively distributed to 10 deliberately selected key professionals working in the Flood Protection Programme. We identified key professionals through judgement sampling. Participants were asked to respond to four to seven propositions per principle, tailored to the context of the Flood Protection Programme. In the final part of the survey, respondents were asked to reflect on the connections between the principles and the outcomes of effectiveness, efficiency, and trust and engagement for the programme.

The survey was completed by five (groups of) experts (a 50% response rate). We recognize that this is a small number of respondents, but given (1) the stature of the respondents and (2) the fact that the survey was merely intended as a first step in a much larger research design, we argue that the results are relevant input for subsequent steps. Next, the outcomes of the survey were discussed in a focus group with executives of four regional water authorities. The discussion in the focus group, with the unique property that participants could not only respond to the researcher's questions but also to react to each other, focused on the provisional results of the assessment of the programme.

In Step 3, we validated the results of the survey and focus group in semi-structured interviews that were audiotaped, transcribed or summarized, and subjected to qualitative analysis. Interviews were conducted with eight high-level experts, again selected by judgement sampling. These experts ranged from a professor in flood risk management to directors and governors in the national Ministry of Infrastructure and Environment, the National Delta Programme, a province, and regional water authorities, to represent all relevant stakes and perspectives within (the context of) the programme. In addition, preliminary results were reviewed from a legal perspective to assess how the preliminary findings related to legal and regulatory concepts, requirements and frameworks.

Steps 2 and 3 can clearly be seen as the 'investigative' part of the assessment, executed by experts with the explicit aim to come to an independent judgement of the programme and performance on the principles. Step 4, on the other hand, consisted of a 'learning table' session, which was organized to reflect on the main outcomes of the assessment, the practical value of the principles for water governance practitioners, and most importantly, to generate lessons on improving the water governance of the programme. In preparing the learning table, we prepared a memo which addressed the main insights and questions of participants in the assessment (per principle). The memo was sent in advance to 15 participants, who were a mix of directors of the Flood Protection Programme, the ministry, regional water authorities, and (independent) experts in water governance.

Results of learning assessment of the Flood Protection Programme

Survey and focus group

Overall, survey respondents agreed that the governance system of the Flood Protection Programme was effective and trustworthy (i.e. legitimate); they partly agreed on its

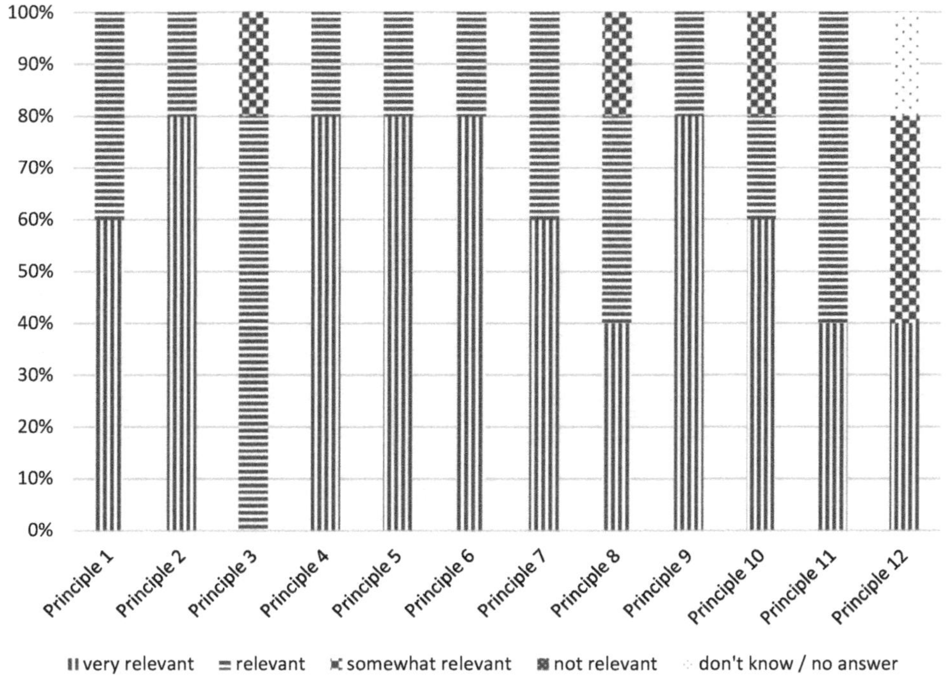

Figure 3. Survey scores on relevance of the 12 OECD principles to the Dutch Flood Protection Programme.

efficiency. Figure 3 shows that respondents considered all the principles to be relevant or very relevant to the Flood Protection Programme. Principle 12 (monitoring and evaluation), showed the weakest score on relevance, but still 40% considered it very relevant and 40% partly relevant.

Figure 4 depicts the extent to which survey respondents considered the programme to be performing in accordance with the principles. There is quite some variance in the perceived performance of the programme. The data suggest that the programme achieves four principles well: Principles 6 (financing), 7 (regulatory frameworks), 8 (innovative governance), and 9 (integrity). The programme fairly achieves Principles 1 (clear roles) and 2 (appropriate scales). The programme partly achieves five principles: Principles 3 (policy coherence), 4 (capacity), 5 (data), 10 (stakeholder engagement), and 11 (trade-offs). Principle 12 (monitoring and evaluation) was considered partly achieved, but was also perceived by the respondents as difficult to assess. In sum, respondents considered that the programme has achieved six principles (1, 2, 6, 7, 8, 9), and partly achieved the other six (3, 4, 5, 10, 11, 12).

The focus group participants agreed to a large extent with these survey results. More important, the focus group participants provided more insight into the effectiveness, efficiency and legitimacy scores of the survey. Despite their agreement with the survey results, they argued that, at this stage, definite conclusions on the effectiveness and efficiency of the programme were impossible to draw, as the programme has not yet resulted in the physical execution of dike-reinforcement projects. They moreover stressed that legitimacy may become a major issue under the Water Act, a new

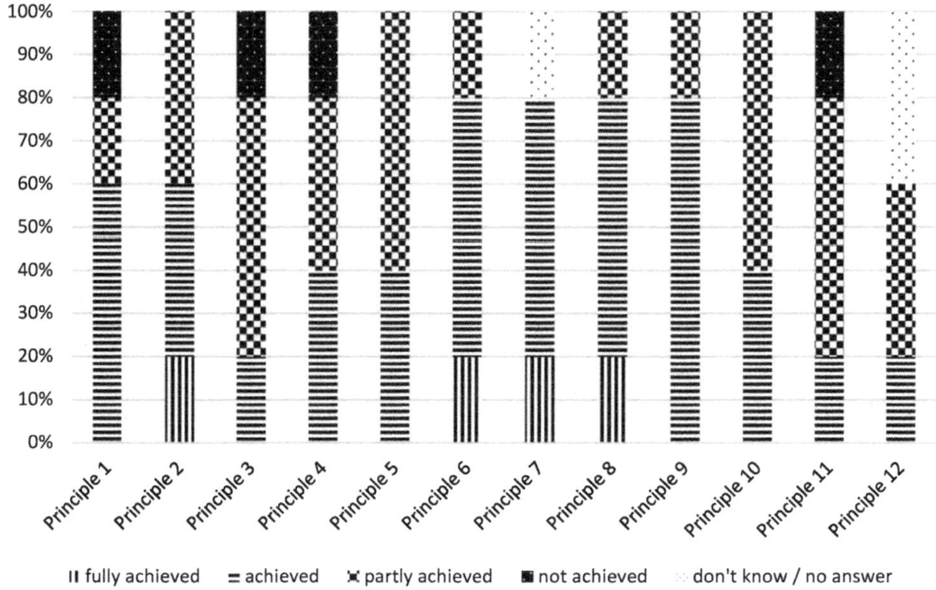

Figure 4. Survey scores on extent to which the Dutch Flood Protection Programme achieves the 12 OECD principles.

regulatory framework (explained below, under Principle 7). The participants shared concerns with five principles:

- *Principle 4, Capacity.* The capacity to implement the programme and to execute the dike-reinforcement projects is likely to become increasingly problematic, as the workload will grow, while too few qualified workers will be available.
- *Principle 5, data and information.* The data and information that regional water authorities submit to the Flood Protection Programme are not always consistent and comparable. As a result, it is difficult to compare the priority of different projects, which is an important starting point for the programme. It may also create instability for the programme, as the scope of projects is very changeable.
- *Principle 7, regulatory frameworks.* A new regulatory framework has recently been adopted in the Water Act, namely a risk-based approach, and highly complex security standards in Dutch flood risk management. This new framework complicates the task of the regional water authorities, because it is not yet clear what the consequences will be in terms of future required dike reinforcements.
- *Principle 8, innovative governance.* Good progress has been made in innovative governance, for instance by initiating specific research and pilot projects on innovative dike-reinforcement techniques that benefited several dike-reinforcement projects. However, this principle remains a major point of attention as innovative techniques should contribute to cost reduction and thus to efficiency, but have not yet proven effective.

- *Principle 9, integrity and transparency.* The programme is a multi-billion-euro programme based on a single fund that several parties contribute to. However, the programme's benefits for each party can vary widely. Respondents stress that integrity should therefore be a major point of attention throughout the programme's course. At the same time they acknowledge that this issue receives relatively little attention in the programme and in the Netherlands' policy debate in general.

Interviews

The interviewees agreed to a very large extent with the survey scores and concerns raised in the focus group. Nonetheless, some contrasting opinions were also voiced. For instance, in relation to Principle 1 (clear roles), an interviewee criticized the allocation of responsibilities between the national and regional water authorities. Although the programme is a joint effort of national and regional water authorities, for some parties the granting of subsidies appears to be merely the task of the national water agency (Rijkswaterstaat). In addition, in relation to Principle 3, some interviewees consider policy coherence, or the incorporation of additional interests and functions into dike-reinforcement projects, a responsibility of the programme, whereas other interviewees see it as an add-on, not a hard requirement. For instance, expansion of a harbour or new bike paths could be fitted into a dike-reinforcement project, thus serving the interests of local actors.

The validation of the survey, focus group and interview findings did not lead to major contrasting results. At the same time it must be acknowledged that perspectives differ on how the responsibilities are or should be allocated between the national and regional water authorities. There are also concerns about the clarity of the new regulatory framework for legal safety standards and the further integration of legislation concerning the living environment.

Learning table

At the start of the learning table session, the findings of the assessment (survey, focus group, interviews) were presented in combination with a set of questions for reflection and learning. These questions, summarized in Table 2, addressed the concerns and critical remarks made by participants in the focus group or during the interviews.

After discussing the main insights, discussions were held on topics related to six principles, resulting in a set of lessons to be learned, summarized below.

Principle 2, appropriate scales. At the national level, not all long-term objectives for flood risk management have been set clearly. This leads to uncertainty in the requirements for separate reinforcement projects. For instance, decisions on extra floodplain restorations or alterations in river bifurcations may affect water levels during floods at a greater scale than a single dike-reinforcement project, thus impacting the goals of separate reinforcement projects.

Principle 3, policy coherence. Generating and maintaining policy coherence is the responsibility of the regional water authorities, who oversee their own dike-

Table 2. Questions formulated for participants in the learning table to stimulate reflection and learning.

OECD Water Governance Principle	Questions
(1) Roles and responsibilities	How to keep the alliance of national and regional water authorities operational? Are responsibilities rightly allocated between national and regional water authorities?
(2) Appropriate scales	How to organize commitment at the regional level to define strategies at the appropriate scale?
(3) Policy coherence	How to come to a structural integration with other policy and land-use functions in dike-reinforcement projects? Should the Flood Protection Programme take more initiative to seek connections with spatial planning?
(4) Capacity	How to prevent capacity shortages? Through more collaboration between regional water authorities, or by carefully planning large, complex projects over time?
(5) Data and information	Is it useful and needed to better organize the flow of data and information from regional water authorities to the Flood Protection Programme?
(6) Financing	Are more and improved financial incentives needed to increase the efficiency of the Flood Protection Programme?
(7) Regulatory frameworks	How to make the complex new regulatory framework of risk-based calculations practically applicable? How to help each other?
(8) Innovative governance	How can the regional water authorities be facilitated to increasingly develop and apply innovative techniques?
(9) Integrity and transparency	Should more attention be paid to risks of corruption (especially in relation to constructors)?
(10) Stakeholder engagement	To what extent are people in society satisfied with the way they can participate in the Flood Protection Programme? Are expectations fulfilled, and do the provisions for participation meet the expectations of stakeholders?
(11) Equity	To what extent is the Flood Protection Programme only responsible for execution of previous (political) decisions, or should trade-offs and equity issues between areas and users be questioned?
(12) Monitoring and evaluation	Which changes in the governance system of the Flood Protection Programme would be useful, and which experiences/examples may be a source of inspiration?

reinforcement projects in their context. When a (regional) project team tries to come to a more integrative project, the Flood Protection Programme does offer extra time to make and shape a dike reinforcement project in which interests of multiple local organizations are integrated. Regional executives have a crucial role in that respect, as they should seek (and know of) opportunities for collaboration with other local organizations and interests.

Principle 4, capacity. In the learning table session, the capacity shortages of regional water authorities were considered more severe than discussed so far in the assessment. The authorities lack the personnel with the necessary competences for complex projects. People with a background in contract management, project management or geo-engineering are especially needed. Sharing capacity between regional water authorities (e.g. pooling of employees) is regarded a fruitful strategy. Also, the recognition of technology in the water sector should increase, both in the water authorities and in relevant educational institutions.

Principle 8, innovative governance. To achieve the Flood Protection Programme's ambitions for cost reduction, the programme is highly dependent on a large-scale application of innovative techniques and approaches in dike reinforcement. To achieve this, four aspects were deemed crucial: continuance of the collaboration in the alliance between national and regional water authorities; upscaling of innovations from potential technologies to agreed-on technologies with high potential for cost reduction, which will then be subsidized; sharing of best-case practices that reduced costs or enabled rapid progress in a project; and collaboration with private constructors.

Principle 9, integrity and transparency. This topic deserves much more attention in the Flood Protection Programme than it has received so far. More attention should be devoted to integrity and transparency to ensure that the water sector is as secure as possible against corruption and fraud. Possible risks of corruption should be mapped in the context of the Flood Protection Programme, for instance in the development and application of innovative technologies in which private companies are involved early on, before projects and technologies are tendered.

Principle 10, stakeholder engagement. The Flood Protection Programme has recently been reformed (as explained in the first half of the paper). This was done in close collaboration between the ministry, the National Water Agency (Rijkswaterstaat) and the regional water authorities. However, stakeholders from outside the flood protection community, such as provinces and universities and other knowledge institutes, could be involved more, to obtain their feedback on the setup and decisions made by the Flood Protection Programme.

Functions and conditions for a practical learning assessment

The case analysis showed that the learning assessment of the Flood Protection Programme had practical value, as the learning table session generated six relevant and practical lessons to enhance the programme's governance system. As Table 3 shows, the principles in relation to which lessons were learned did not surface in each step of the learning assessment. Not only does this illuminate the challenging nature of assessing practices within water governance systems, it also confirms the relevance of applying a variety of methods in the learning assessment. The survey allowed participants to give their opinion on all principles; the focus group enabled collective discussion of the meaning of the principles and the performance of the programme; and the interviews provided a 'safe environment' for professionals to

Table 3. Principles highlighted in the assessment to improve the performance of the Dutch Flood Protection Programme, either because they were partly achieved (survey), or because participants shared concerns (focus group, interviews) and lessons for improvement were drawn (learning table).

OECD Water Governance Principles	1	2	3	4	5	6	7	8	9	10	11	12
Survey			■	■	■					■	■	■
Focus group				■	■		■	■	■			
Interviews	■		■	■	■		■	■	■			
Learning table		■	■	■				■	■	■		

voice their concerns. The learning table session provided a platform for synthesis and learning on coping with key water governance challenges.

The OECD Water Governance Principles had multiple functions throughout the learning assessment. First, the principles enable discussions of the water governance system and enhance understanding of the systems' functioning. As exemplified by discussions of policy coherence during the interviews and the learning table, the principles can help identify differing perspectives on what a governance system can or cannot, should or should not do. Second, the principles highlight a broad spectrum of requirements for good water governance and are thus helpful in identifying 'hidden' risks and challenges that are currently not on the agenda of directors in the Flood Protection Programme, as has been illustrated by outcomes about integrity in the survey and the learning table session. Third, the principles can stimulate reflection on actual governance challenges, as joint discussion thereof can be helpful to identify, specify and prioritize key challenges. For instance, the issue of capacity appears to be more urgent than expected, and the seriousness of successful innovative governance for efficient governance of the programme was rediscussed during the learning table session. Lastly, the fourth function is informed action to respond to the challenges revealed by the governance assessment. Although capacity had already received major attention in the programme, the assessment reconfirmed the severity of the programme's limited capacity and the need to put more effort into sharing knowledge between organizations and persons, and cooperating with universities and universities of applied science to interest future employees.

Several conditions were present during the learning assessment, stimulating these four functions to come to full bloom. First, the principles were translated to the context of the Flood Protection Programme in each step of the assessment. The translation helped make the principles meaningful to the participants of the survey, focus group, interviews and learning table. Second, the assessment was undertaken in a period without major policy debates or controversies that could politicize the process and outcomes of the assessment. The absence of political pressure created a safe environment in which people could more openly discuss the strengths and weaknesses of the Flood Protection Programme's water governance. Third, a diversity of stakeholders was involved in the interviews and the learning table. They mostly represented the interests and insights of the water authorities, but these were supplemented by a provincial governor and water governance academics. Through involvement of these key stakeholders, the interests and perspectives of those parties that constitute the water governance system were involved in the assessment.

Implications and conclusions

The aim of this article was to explore the practical value of the OECD Water Governance Principles in assessing water governance practices, through developing a learning assessment that generates lessons to enhance the performance of water governance systems. This article revealed how the principles had practical value in generating lessons to strengthen the effectiveness, efficiency and legitimacy of the Dutch Flood Protection Programme. In addition, the assessment revealed functions of the OECD Water Governance Principles in enabling and stimulating learning to strengthen a water governance system. These functions are (1) to enhance understanding of the water governance system, (2) to reform the agenda, (3) to reflect and set priorities, and (4) to inform action. These functions cover to a large extent the ones discussed in the theoretical section on reflexive learning in governance assessments, except for changes in attitude and governance systems (Ison et al., 2007; Pahl-Wostl et al., 2007). Real changes in attitude and governance systems have not been observed, as the research project ended after the learning table. Therefore, the learning assessment has mainly pointed out to key parties involved in the programme what they consider the most important water governance challenges, and how these could be addressed. But this includes recommendations for a change in attitude or the governance system. As most of the learning functions were in place, it can be concluded that the OECD Water Governance Principles can be used as an effective instrument to contribute to good water governance.

Reflection

Although this article advances the understanding of how to apply the principles in a meaningful water governance assessment, there are some limitations in the way the learning assessment was conducted. The assessment had primarily an internal focus, which involved key actors in the specific water governance system of the Flood Protection Programme. This resulted in a learning assessment that was conducted in a safe, depoliticized environment, but also in a narrow dialogue within one epistemic community, which shares core beliefs, policies and claims to authoritative knowledge (Haas, 1992). Except for one provincial governor, critical voices from outside the flood protection community were not involved. This limited variety in stakeholder involvement reduces the empirical value of the assessment outcomes. If people from outside the Dutch flood protection community were involved (e.g. community representatives, NGOs, municipalities, private parties), a more representative assessment of the programme would have been conducted. Such a more inclusive assessment would be in line with the bottom-up and inclusive decision making advocated by the OECD (2015) and the suggestion of Bressers et al. (2016) to involve potential critics in water governance assessments. Critical outsiders should thus be involved through interviews or focus groups to reflect on assessment outcomes, and when possible participate in learning tables. If the latter is not possible, it is the responsibility of the assessment team to present critical outsiders' views in the learning table. Though the reflexive assessment generated six lessons learned, it was not possible to assess how these lessons will be incorporated in the governance system. Thus, it could not be determined whether learning was transformed into action. This would require a post-evaluation or another assessment in 3–5 years with involvement from a broader variety of stakeholders.

The OECD principles can become an effective instrument to enhance policy coordination and can contribute to good water governance. However, to realize their full potential the principles should not be used as merely an (internal or external) auditing tool. It is very important to find ways to contextualize the principles each time they are applied, to focus on actual practices instead of assessing governance structures, and to make them relevant to people in the water governance system to be assessed. Furthermore, it is essential that the principles are not only used to gather information to be able to give a judgment, but that this information is deliberately used to get new issues on the agenda, to clarify ambiguities, to facilitate frame reflection and social learning, and to spur action. Our method can be seen as a first step towards using the principles in such a way. In addition, periodic assessment is needed to move beyond the one-off learning events, also adhering to notions of social learning being situated in wider social units beyond the persons directly involved in an assessment (Reed et al., 2010).

The potential applicability of the OECD Water Governance Principles and water governance assessments is enormous. With the rise of water management institutions in the 1700s–1900s under the scientific paradigm of water management (Hassan, 2011), monitoring water in all its aspects has received much attention, because 'you cannot manage what you do not measure'. Yet in recent decades a new insight emerged: that many water-related problems are problems of governance. To build on that new insight, it can be argued, water governance assessments should receive more attention worldwide, as evaluation and learning about the performance of water governance systems can enhance countries' capacity to cope with current or upcoming water crises. Globally operating intergovernmental organizations (e.g. the UN World Water Assessment Programme, the Global Water Partnership, the OECD) could set the agenda for reflexive, action-informed water governance assessments.

Note

1. The OECD Water Governance Initiative is a policy forum where public, private and not-for-profit organizations meet in support of better governance for the water sector. The 140-plus members reflect the diversity of organizations that are concerned with water governance, ranging from national governments and other water authorities (regional, local, river basins) to international organizations, NGOs, financial institutions, research institutes and universities.

Acknowledgements

We thank the participants for sharing their views on the water governance of the Flood Protection Programme. Many thanks to Luuk Claesens, who assisted in data collection and analysis. We greatly appreciate the constructive feedback of two anonymous reviewers; their comments helped clarify the strengths and limits of this study.

Disclosure statement

No potential conflict of interest was reported by the authors.

Funding

This work was supported by Deltares under the Delta Governance research programme.

References

Akhmouch, A., & Clavreul, D. (2016). Stakeholder engagement for inclusive water governance: "Practicing what we preach" with the OECD water governance initiative. *Water*, *8*(5), 204.

Argyris, D., & Schön, D. (1978). *Organizational learning: A theory of action perspective*. Reading: Addison-Wesley.

Asian Development Bank. (2016). *Asian water development outlook 2016: Strengthening water in Asia and the pacific*. Mandaluyong City, Philipines: Author.

Bressers, H., Bressers, N., Kuks, S., & Larrue, C. (2016). The governance assessment tool and its use. In H. Bressers, N. Bressers, & C. Larrue (Eds.), *Governance for drought resilience: Land and water drought management in Europe*. Basel: SpringerNature.

Davies, I. C. (1999). Evaluation and performance management in government. *Evaluation*, *5*(2), 150–159.

Driessen, P. P. J., Hegger, D. L. T., Bakker, M. H. N., van Rijswick, H. F. M. W., & Kundzewicz, Z. W. (2016). Toward more resilient flood risk governance. *Ecology and Society*, *21*(4).

Edelenbos, J., & van Buuren, A. (2005). The learning evaluation: A theoretical and empirical exploration. *Evaluation Review*, *29*(6), 591–612.

Gupta, J., Termeer, C., Klostermann, J., Meijerink, S., Van Den Brink, M., Jong, P., ... Bergsma, E. (2010). The Adaptive Capacity Wheel: A method to assess the inherent characteristics of institutions to enable the adaptive capacity of society. *Environmental Science & Policy*, *13*(6), 459–471.

Haas, P. M. (1992). Introduction: Epistemic communities and international policy coordination. *International Organization*, *46*(01), 1–35.

Hajer, M., Nilsson, M., Raworth, K., Bakker, P., Berkhout, F., de Boer, Y., ... Kok, M. (2015). Beyond cockpit-ism: Four insights to enhance the transformative potential of the sustainable development goals. *Sustainability*, *7*(2), 1651–1660.

Hassan, F. (2011). *Water history for our times*. Paris: UNESCO Publishing.

Havekes, H., Hofstra, M., Kerk, A., Teeuwen, V. D., Cleef, R. V., & Oosterloo, K. (2016). *Building blocks for good water governance*. The Hague: Water Governance Centre.

Hill, M., & Hupe, P. (2002). *Implementing public policy: Governance in theory and practice*. London: Sage Publications Ltd.

Huitema, D., Mostert, E., Egas, W., Moellenkamp, S., Pahl-Wostl, C., & Yalcin, R. (2009). Adaptive water governance: assessing the institutional prescriptions of adaptive (Co-)management from a governance perspective and defining a research Agenda. *Ecology and Society*, *14*(1).

Ison, R., Röling, N., & Watson, D. (2007). Challenges to science and society in the sustainable management and use of water: Investigating the role of social learning. *Environmental Science & Policy*, *10*(6), 499–511.

Jessop, B. (2003). Governance and metagovernance: On reflexivity, requisite variety, and requisite irony. In H. P. Bang (Ed.), *Governance as social and political communication* (pp. 142–172). Manchester: Manchester University Press.

Jorissen, R., Kraaij, E., & Tromp, E. (2016). Dutch flood protection policy and measures based on risk assessment. In E3S Web of Conferences (Ed.), *3rd European Conference on Flood Risk Management (FLOODrisk 2016)*. Lyon.

Keen, M., Brown, V. A., & Dyball, R. (2005). Social learning: A new approach to environmental management. In M. Keen, V. A. Brown, & R. Dyball (Eds.), *Social learning in environmental management: Towards a sustainable future*. Abingdon: Earthscan.

Kind, J. M. (2014). Economically efficient flood protection standards for the Netherlands. *Journal of Flood Risk Management*, *7*(2), 103–117.

Lautze, J., De Silva, S., Giordano, M., & Sanford, L. (2011). Putting the cart before the horse: Water governance and IWRM. *Natural Resources Forum*, *35*(1), 1–8.

Mezirow, J. (1995). Transformation theory of adult learning. In M. Welton (Ed.), *In defense of the lifeworld: Critical perspectives on adult learning*. New York, NY: State University of New York Press.

OECD. (2015). *OECD principles on water governance*. Paris: Author.

Pahl-Wostl, C. (2009). A conceptual framework for analysing adaptive capacity and multi-level learning processes in resource governance regimes. *Global Environmental Change*, *19*(3), 354–365.

Pahl-Wostl, C., Craps, M., Dewulf, A., Mostert, E., Tabara, D., & Taillieu, T. (2007). Social learning and water resources management. *Ecology and Society*, *12*(2).

Reed, M., Evely, A. C., Cundill, G., Fazey, I. R. A., Glass, J., Laing, A., ... Stringer, L. C. (2010). What is social learning? *Ecology and Society*, *15*(4), r1.

Romme, A., & Van Witteloostuijn, A. (1999). Circular organizing and triple loop learning. *Journal of Organizational Change Management*, *12*(5), 439–454.

Royse, D., Thyer, B. A., & Padgett, D. K. (2010). *Program evaluation. An introduction*. Belmont: Wadsworth Cengage Learning.

Sabel, C. F. (1994). Learning by monitoring: The institutions of economic development. In N. Smelser & R. Swedberg (Eds.), *Handbook of economic sociology*. Princton: Russel Sage and Princeton University Press.

Seijger, C., Dewulf, G., Otter, H., & Van Tatenhove, J. (2013). Understanding interactive knowledge development in coastal projects. *Environmental Science and Policy*, *29*, 103–114. Retrieved from http://www.scopus.com/inward/record.url?eid=2-s2.0-84875035183&partnerID=40&md5=151664bfd4578e08b2ace71cd34e90ac

Teisman, G., van Buuren, A., Edelenbos, J., & Warner, J. (2013). Water governance: Facing the limits of managerialism, determinaism, water-centricity and technocratic problem-solving. *International Journal of Water Governance*, *1*(1), 1–11.

Termeer, C., Dewulf, A., van Rijswick, H. F. M. W., van Buuren, A., Huitema, D., Meijerink, S., ... Wiering, M. (2011). The regional governance of climate adaptation: A framework for developing legitimate, effective, and resilient governance arrangements. *Climate Law*, *2*(2), 159–179.

Thieken, A., Kienzler, S., Kreibich, H., Kuhlicke, C., Kunz, M., Mühr, B., ... Schröter, K. (2016). Review of the flood risk management system in Germany after the major flood in 2013. *Ecology and Society*, *21*(2).

United Nations World Water Assessment Programme. (2015). *The United Nations world water development report: Water for a sutainable world*. Paris: Author.

van Rijswick, H. F. M. W., & Havekes, H. (2012). *European and Dutch water law*. Groningen: Europe Law Publishing.

van Rijswick, M., Edelenbos, J., Hellegers, P., Kok, M., & Kuks, S. (2014). Ten building blocks for sustainable water governance: An integrated method to assess the governance of water. *Water International*, *39*(5), 725–742.

World Economic Forum. (2016). *The global risks report 2016: 11th edition*. Geneva: Author.

World Water Assessment Programme. (2003). *The united nations world water development report: Water for people water for life*. Paris: UNESCO.

The evolution of water governance in France from the 1960s: disputes as major drivers for radical changes within a consensual framework

Marine Colon, Sophie Richard and Pierre-Alain Roche

ABSTRACT

This paper provides a synthetic presentation of French water governance and its evolution since the 1960s. Through this French experience, it discusses the Organisation for Economic Co-operation and Development (OECD) water governance cycle showing disputes as the main drivers of change. France has been a pioneer in introducing water river basin management some 50 years ago. It is also noted for its water services management by local authorities, leaving a significant role to private and public companies. But French water governance has not been frozen since the 1960s and continues to change radically within a framework based upon its unique history.

Introduction

There is now a consensus about the fact that water management not only requires the right techniques and sufficient finance but also above all an 'enabling' governance system (e.g., Gurría, 2009). The Organisation for Economic Co-operation and Development (OECD) has taken a significant step towards such recognition by adopting 'governance principles' designed to guide governments in their quest to improve their water-governance framework. These principles are built on best practices and have been prepared by the Water Governance Initiative, which was cofounded by ASTEE[1] and led by the OECD. Their goal is for 'effective, efficient and inclusive water policies in a shared responsibility with the broader range of stakeholders'.[2] However, these guidelines say little about how governance systems adjust with time to changing environments, challenges and societal expectations.

France's water-governance system may be seen as a stable and even dormant framework built upon two pillars: integrated water resource management at river basin scale (introduced by the first water law in 1964) and the public–private partnership model of local water and sanitation services which extended from urban to rural areas during the course of the 1960s/70s. Although this framework remains consensual, the objectives of water policies and the tools for their implementation constantly evolve. We still need to understand precisely what are the drivers of change and what directions this governance system is taking.

The purpose of this paper is threefold. First, it analyses how water governance has developed in France. We will detail the three characteristics of such change: disputes as one of the change drivers, resulting in more incremental than radical evolutions due to a form of stable consensus concerning the water-governance framework. Second, it provides an overview of recent changes introduced in France's water-governance system about which surprisingly little appears in the literature (Barbier, 2015; Barraqué & Laigneau, 2017). Third, and more broadly, it sets out to discuss the water-governance cycle suggested by the OECD. Water governance is understood here as including both water resource management and water service management.

The paper is based upon a literature review, and the various professional experiences of the authors within the water sector, as water professionals, researchers and teachers involved in policy-making processes, members of the OECD governance initiative, experts in water utility management and water governance. Therefore, it provides an in-depth understanding of the French context, although inevitably carrying with it a very French and expert point of view.

This paper is structured as follows. The first and second sections analyse the evolution of the French water-governance system referring to the OECD governance principles, which are divided into water services governance and water resource governance. The principles are used as a frame in this analysis and help to identify how far these principles are followed, how they are implemented and which are guiding current reforms. The third section details change processes to highlight the three characteristics of the change dynamic: disputes and incremental change within a consensual framework. The final section discusses the OECD water-cycle framework and puts forward perspectives.

The evolution of France's water services governance system[3]

Water governance is closely connected to governance in the public sphere. Prerequisites on French governance are set out in Box 1. While water and sanitation services' governance relies on local authorities, water resource governance required setting up specific and innovative structures.

The framework

France has a unique model of water and sanitation service governance (Roche, Guerber, Nicol, & Simoni, 2016) deriving from its history (Box 2). It is based on five major elements: (1) duality as both public and commercial services; (2) local management (decentralization); (3) public management with private participation; (4) fragmented state regulation; and (5) funding by users. The main stakeholders are set out in Figure 1.

- Water and sanitation services are by law recognized as both public and commercial services. Being public, they must comply with French public service core principles: continuity of service (continuous service delivery), and equality of treatment of users and mutability (infrastructures and service delivery to comply

Box 1. French state and local authorities governance frame.

France has a long history of the centralization of power. However, since the French Revolution of the late 18th century, governance has been divided into two distinct spheres: state and local authorities. There are three types of local authorities: municipalities, *départements* and regions. One peculiarity of France is the vast number of municipalities – more than 36,000. The country is divided into 101 *départements* and regions whose numbers recently decreased from 27 (22 in metropolitan France – MF) to 18 (12 in MF). While municipalities and *départements* were established with the Revolution, regions governed by locally elected representatives were only created in 1982, with the first decentralization statute. The state used to take part in decisions taken by municipalities, but this law restricted its role to controlling legal compliance and assisting and contributing to finance local public actions. The state consists of ministries and public bodies at the national level, with local representations. Heads of ministries belong to the government. Local authorities are headed by elected representatives. Financially, local authorities levy local taxes and receive grants from the central state.

Box 2. Origins of the French model of water and sanitation services' management.

In France, as in many other industrialized countries, drinking water and sanitation network infrastructures developed in the second half of the 19th century. The awareness that these infrastructures were crucial to public health came soon after the plague epidemics that afflicted Europe (Paris in 1833 and 1849) and scientific pioneering discoveries (Louis Pasteur, Florence Nightingale, Robert Koch etc.). Municipalities had been in charge of providing water and sanitation services since the French Revolution, but always lacked the funding to implant their allocated role (Pezon, 1999). To encourage infrastructure development, the government carried out a large urbanization refurbishment programme in Paris whereby private entrepreneurs invested in water and sanitation networks (Crespi-Reghizzi, 2014). In addition, two companies were founded: the Compagnie Générale des Eaux in 1843 (later becoming VEOLIA eau) and the Société Lyonnaise des Eaux et de l'Eclairage in 1867 (subsequently becoming Suez), which took over the development of services in a number of major cities such as Lyon and Bordeaux. From there, two models of service management coexisted in France: concession and public management ('régie'). The legal frameworks to clarify the roles and duties of the state, municipalities, private companies and users (tariff policy, controls, investment) were then set out at various stages during the course of the 20th century.

with laws,[4] norms and available technologies at any time). Being commercial, they are entitled to sell services to users.

- Water and sanitation services' management is indeed local since the 36,000 French municipalities are responsible for water and sanitation services. This means they fund and own the infrastructures and are responsible for service delivery, for which they may contract with private or public companies.
- Municipalities may contract with private (or public) companies for operation, maintenance and commercial services (*'délégation de service public'*) up to infrastructure investment (concession model) (Figure 2). Delegation is in France more widespread than anywhere else in the world (Clark & Mondello, 1999). Around 60% of the French population drinks tap water from services operated by a private operator (ONEMA, 2016). This model of delegation developed during the post-Second World War rebuilding of France, with rapid infrastructure development and little capacity to handle them in many very small municipalities. The state played a major role in developing services and rolling out the delegation model.
- Even if municipalities are in the front line to handle water and sanitation utilities, the state has nevertheless always played a key role in two respects: sector development and regulation (including policy-making). While its role in sector development may be considered as transitory (see later developments), the state is the

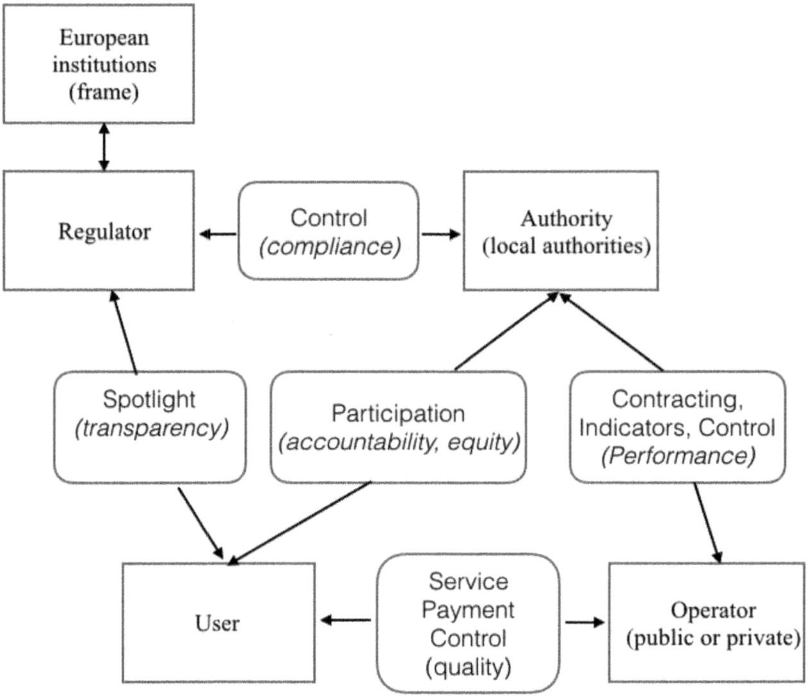

Figure 1. Stakeholders in France's water service governance.
Source: Adapted from Roche, Colas-Berlcour, Vial, and Tandonnet (2016).

Public management	- direct management by local authority's governance structure - specific budget (funded by water bills) - headed by a civil servant - employment contracts under private law
Public local enterprise	- financial autonomy - legal existence - headed by a civil servant accountable to local authority - employees under private law - under contract without bidding and no deadline - may have private shareholders
Private enterprise	- majority of private shareholders - legal existence - head and employees under private law - under contract through bidding and time limit

Figure 2. Model of water services' management in France.

regulator. Regulation is not implemented by any formal regulatory agency as in England and Wales with Ofwat, the Water Services Regulation Authority. Regulation is overseen by a number of state services, making state regulation fragmented (Canneva, 2012). Central and local state services ensure the respect of health, environmental, accounting and competition rules.

- Users contribute to water and sanitation services management by funding them through the paying of water bills. Water bills fund water and sanitation services management (investment, operations and maintenance (O&M), and commercial services), and water agencies through user fees. In France, this financing model is known as 'water pays for water'. Local authorities in charge of water and sanitation services manage a specific budget in which income has to come from the activity of the service (water sales) and expenses have to finance either operation, maintenance and investment dedicated to the service. Water and sanitation services are entitled to resort to bank loans and public funds to finance investments. Social tariffs to ease access to poorer households have been tried in various cities (Lille, Nantes, Grenoble, Paris).

In 2013, 61% of the French population was supplied with drinking water by a private enterprise under contract with a local authority. This rate is 41% for sanitation services. Three main private companies share 99% of the market shares of concession contracts in France: Veolia, Suez and Saur. A total of 1% of shares is left to smaller independent companies. This water services governance system is characterized by the main responsibilities borne by local authorities, the strong implication of private companies in services' management and the existence of an ancient oligopoly. It is weakened by the large number of small local authorities in charge of water services. A total of 56% of water utilities serve 4% of the population; 10% of water utilities serve more than 50,000 inhabitants each (ONEMA, 2016). These small authorities are typically rural, low staffed, which make them unattractive to private operators, with low investment capacities. The renewable rate of the water network is of 0.58% (ONEMA, 2016), which is low. The responsibility for renewing networks is shared between the local authorities and the public or private operators.

This governance system for water and sanitation services remained relatively stable over the last 50 years. Changes were, however, introduced to improve performance for improved trust and engagement (principle 9 on integrity and transparency of practices, principle 10 on user involvement in decision-making and public information) and effectiveness (principle 4 on strengthening the capacity of responsible authorities).

User involvement, strengthening of local authorities and the new role of the state limited to regulation

During the course of the last half century, the legal framework organizing water and sanitation utility governance has evolved three times: during the mid-1990s with the Sapin law (1993), the Barnier and Mazaud laws (1995), the mid-2000s with the Water Act (2006) and the 2007 state Reform ('*Réforme Générale des Politiques Publiques*'), and then the mid-2010s with the third decentralization laws (NOTRe 2015). The 1990s and 2000s laws tackled the issue of trust and engagement, while the 2007 state Reform and the 2010 decentralization law focused on the capacity of responsible authorities and the role of the state.

In the 1990s three major laws were passed, designed to boost trust and engagement in the sector by imposing specific competition rules, ensuring availability of information to citizens and accountability from private operators. In 1993, the Sapin law introduced competition rules specific to delegation contracts for public services such as water and sanitation. In 1995, the Barnier law obliged water and sanitation services to produce annual reports to inform citizens about key facts and figures. The Mazaud law then obliged private companies to hand over an annual report on their management to responsible authorities. Later, in 2002, a law on local democracy obliged responsible authorities serving more than 10,000 people to involve users in service governance. User committees henceforth had to be consulted over key decisions such as moving from delegated to fully public operations. These laws introduced notions of transparency and integrity, trust and engagement into the sector.

Another set of laws tackled the capacity of responsible authorities to manage water and sanitation services, closely linked to the role of the state. As mentioned previously, since the 1950s the role of the state has gone beyond mere regulation since it contributed to developing and organizing the sector. 'Deconcentrated' state services, known as 'public engineering services', played a major role in governance. Composed of state engineers under the responsibility of either the Ministry of Agriculture or the Ministry of Public Works, these services were organized at the scale of *départements*. This local grounding offered state engineers in public engineering services an understanding of specific local issues, and integration in the web of relationships between local authorities, their contracting companies and any local actors. Public engineering services administered funds for infrastructure development, assisted responsible authorities by designing and managing infrastructure development projects and, subsequently, assisting them in contracting with the private sector to manage new infrastructures. They also worked on the design of new services involving the cooperation between municipalities, in the case of shared water resources, for instance.

This governance model brought with it two major issues. First, water and sanitation services remained too small in scope. In 2013, 56% of the 13,530 French water services served fewer than 1000 people, and only 2% served more than 50,000 people (ONEMA, 2016). Small services rely on external consultancies because they clearly cannot hire permanent expert staff. This lack of autonomy is clearly an obstacle to building the consistent long-term development strategy required for sound management. In addition, a large number of water and sanitation services (34,714 in 2013; ONEMA, 2016) triggers a significant disparity of water tariffs up and down the country, and increases the complexity and cost of regulation. The mere building of a national information system, imposed by the water law in 2006, came across as a most challenging task. The decentralization law of 2015, known as the 'NOTRe' law for 'Nouvelle Organisation Territoriale de la République' (New Territorial Organization of the Republic), addressed the main obstacle to strengthening local authorities: their sheer number and size. The remunicipalization of Parisian water services reminded everyone that local authorities are the key actors in water and sanitation service management (Valdovinos, 2012). This law specified that the responsibility of water and sanitation services should be transferred to groupings of municipalities by 2017. This was part of an earlier rationalization movement taking place in other European countries such as in England and Wales with the 1973 Water

Act and in Italy with the 1994 Galli law (Barraqué, Isnard, Barbier, & Canneva, 2011). This rationalization involves moving from local communes to larger local authorities to take advantage of economies of scale, but also for the better interconnectedness of water networks and improved safety in drinking water provision (Barbier, 2015).

Another issue confronting this management model is the role of the state. As already explained, since the 1960s the state has played a crucial role in organizing the sector with its public engineering services. In the 2000s, this public consultancy service started to be seen as competing with the private consultancy sector. It was also perceived as preventing municipalities from feeling the need to merge to strengthen capacity. In other words, it was not helping to solve the problem of too many and too small utilities. Eventually, the reform of the French state led, in 2008, to the decision to dispense with this activity. The state focused instead on its regulatory role (Barone, Dedieu, & Guérin-Schneider, 2016).

Recent detailed proposals have been made to use the opportunity offered by the process of the concentration of public authorities to reinforce central regulation, in a movement referred to by the authors as 'from moonlight to sunshine regulation' (Roche, Guerber, et al., 2016). The French national water committee ('Comité national de l'eau') and the French government decided at the end of 2016 to implement these proposals progressively, taking into account efforts that municipalities will have to dedicate to rebuilding their services at new scales.

Change in water resource governance: towards a more inclusive policy

A governance framework relying on water policy principles designed to attain consensus

France has a unique water resource governance system based on an integrated water management system. It derives from an extensive but historic regulatory framework combining bottom-up and top-down processes. Three water policy principles apply (Levraut et al., 2013). The first states that 'water is part of the common heritage of the nation'. Water resource management should guarantee adequate water quality and quantity for both human use and environment preservation. The second principle stipulates that 'water use belongs to all' and that 'each individual has the right to access drinking water at an acceptable cost'. The third principle stems from the application of the polluter-pays principle. These three principles are the cornerstones of the current water policy consensual framework combining: (1) an autonomous water policy where water agencies play a central role; (2) a decentralized water resource management at basin scale; (3) state regulation; and (4) participatory processes for conciliation between users and engaging civil society.

The following section considers the role and evolution of this system.

Framework development: water management at basin scale, water user engagement, strengthening of local authorities and the new role for the state

France's water-governance framework celebrated its 50th birthday in 2016. From an external perspective it might well look surprisingly stable. Yet, water governance has experienced significant changes regarding both the regulatory framework and the processes involved. The four major legislative steps that shaped current water governance are as follows:

- In 1964, the first Water Act set up an original institutional and financial system for integrated water resource management at basin scale.
- In 1992, the second Water Act went further regarding innovative governance practices at an appropriate management scale (smaller catchments) and reinforced a multi-stakeholder approach.
- In 2006, the third Water Act integrated principles from the 2000 European Water Framework Directive (WFD). It strengthened water management at basin scale, enhanced monitoring and evaluation as well as trust and engagement. The WFD also induced a paradigm shift towards result-oriented management.
- More recently, the 2014 and 2015 decentralization laws and the 2016 biodiversity law strengthened the role of local authorities with jurisdiction over water resource management and flood prevention, and created new tools to link water issues more effectively to other environmental concerns.

1964 Water Act: the establishment of water agencies

The 1964 Water Act set up the foundations of the current water-governance system. The context was a period of high economic growth, a planned economy and a centralized political system with the state as the sole policy-maker. The law decentralized water governance (Richard, Bouleau, & Barone, 2010). It introduced the concept of river basin management, set up six water agencies and their basin committees. It induced a radical paradigm shift from the management of water integrated in fragmented sectorial policies (irrigation, hydroelectricity, sanitation etc.) towards the management of water at the scale of a hydrographic territory. It laid the basis of environmental regulation in France in a context of rapid economic growth involving the development of highly polluting industries and intensive urbanization of natural areas such as coastal areas (Drobenko & Fromageau, 2015). This law may rightly be considered as the first step in France towards recognizing the importance of environmental issues (Bouleau, 2007).

The issue of water management in France at that time was not the availability of the resource but its quality and conflicts between users (Box 3). This new model of territorial management of water was inspired by financing mechanisms set up on the Rhine river basin in Germany (Barraqué & Laigneau, 2017). Relying on a physical reality (the basin), water agencies blurred administrative boundaries and brought together neighbouring areas and users. They are in charge of financing and planning water policy within the area they

Box 3. Key figures on the demographics and water resources in France.

Population: 1960, 46 million; 2013, 66 million; 2017, nearly 67 million at 106 inhabitant/km^2 (INSEE; https://www.insee.fr/fr/statistiques/1892086?sommaire=1912926)
Annual growth rate: 1960, 1.2%; 2013, 0.5%
Water availability: 191 billion m^3/year; needs in 2012, around 17% (30 billion m^3/year) (Table 1)

Table 1. Volumes of water withdrawn per use in France (m^3/year) (SOeS, 2015).

	Water abstraction/withdrawal, 2012	
Water uses	billion m^3/year	%
Drinking	5.5	18.3
Industry	2.9	9.5
Energy	18.7	62.5
Irrigation	2.9	9.7

Source: SOeS, 2015.

serve. They collect fees from water users. They reallocate funds through subsidies and loans to support studies and infrastructure projects designed to improve water quality and availability. From the 1970s onwards, many local water authorities emerged as a result of bottom-up processes. They were specifically set up by groupings of local authorities to take over water resource management issues at river or catchment scale. As previously mentioned, these innovative catchment or river management authorities cross the traditional administrative public demarcation consisting of the state and local authorities. They are the key stakeholders of the French water resource governance system.

Hence, contracting authorities ('*maîtres d'ouvrage*') are both local authorities, be it at local, district or regional scales, and local water authorities, at catchment or river scales.

Water agencies collect around €2 billion per year (Roche, Guerber, et al., 2016). Around 85% of this sum comes from domestic water users. The budgeted amount for the period 2013–18 is €13.3 billion for preservation (Agences de l'Eau, 2017).

The model of a water agency was meant to implement the 2015 polluter pays principle for the first time.[5] It also set up cross-subsidies and solidarity mechanisms at basin scale between urban and rural areas. Basin committees (so-called 'water parliaments') gather representatives from local authorities, water users and the state. The committees act as forums for negotiation, consultation, guidance and decision-making on water at the territorial level of the catchment, enabling a bringing together of different stakeholders with sometimes competing interests. They are in charge of water-use allocation and defining the policy of their water agency. Regarding planning, water agencies act under the joint responsibilities of the state representative and of the basin committee (Nicolazo, 1993).

1992 Water Act: river basin planning tools, collective decision-making ('concertation')

The second important step towards a more inclusive water policy in France was the 1992 water law. Adopted in a context of decentralization and modernization of state services, it reflects a still-growing awareness of environmental issues, in response to serious accidents or major pollution (e.g., Chernobyl and Sandoz, both 1986). This water law defined water as a common heritage of the nation. It introduced new tools for a planned and participatory water management at basin scale. It allowed new forms of public action through local participatory mechanisms. Processes of preparing planning tools such as master plans (SDAGE, at the scale of the river basin) or plans for water management (SAGE, at the scale of a smaller catchment area) were opened to representatives of water users and civil society, through basin committees and local water commissions (catchment areas). At the same time, the act created conditions for greater control of the state. Both planning tools (SDAGE and SAGE) and following activities and operations must still seek approval from the state (Peyrou & Roche, 2006; Richard & Rieu, 2009). At catchment scale, SAGEs are managed by water authorities.

2006 Water Act: new norms from the 2000 European Water Framework Directive

The building of a comprehensive European legislative framework was clearly structured by models promoted at international level such as integrated water resources management (IWRM) (Box 4 has a short history of the European Union policy on water). Although this concept emerged in the 1970s, awareness grew among European institutions in the 1990s (Molle, 2012). The 2000 European WFD resulted from the recognition of the fragmented nature of European water policy, the need to account jointly for the different components of water (quantity, quality, ecology) and for the diversity of water uses (including the environment per se) (Kallis & Butler, 2001). Designed to enhance the effectiveness of water policy, it builds on four key components: ecosystem-based objectives (ultimate objective of a good overall quality of all waters) combined with economic requirements; the planning of water management at river basin scale; cost recovery for water services; and the participation of all stakeholders including civil society as a cornerstone for building a water democracy. According to Salles (2009), participation sets out to build mutual responsibilities between actors. At least it makes issues more transparent whilst at the same time making actors accountable for the environmental results achieved. It introduced a shift towards result-oriented management, with water quality targets to be met by 2015, 2021 or, ultimately, 2027.

With the 2006 Water Act, the European WFD was transposed into French law. It created ONEMA,[6] a central office for freshwaters, to enhance links between water cycle and water service management. It also sought to improve consistency between European guidelines and decentralized implementation at basin scale (Levraut et al., 2013). ONEMA's creation looks like a re-centralization as regards the monitoring and surveillance of water quality, which is a major issue in the context of achieving good status or good potential as advocated by the WFD (Richard et al., 2010).

The enforcement of the European WFD was not such an administrative issue in France regarding the governance framework, since water agencies already had jurisdiction over major watersheds. The WFD in fact strengthened the legitimacy of existing institutions. It also contributed to new forms of governance and collective actions across different territorial scales and decision-making levels. However, processes of setting and reaching objectives of water quality ('good status') and implementing cost-recovery policy have been more challenging for France as they required adjustments (Richard et al., 2010; Roche, 2002). The WFD enhanced monitoring and evaluation. The 'good status' indicators have become the main criteria for assessing the soundness and effectiveness of water management in a given river basin. It gave a new impulse to ecosystem research (Roche, 2005). At the crossroads between democracy and policy efficiency, participation required that transparent information and decision-making at the scale of river basins became the norm.

Box 4. Timeline of European Union water policy. Sources: Kallis and Butler (2001); Bouleau and Richard (2011).

Environmental thinking came to the fore in the 1960s. It began to materialize in concrete form in the 1970s, building on international events and publications (Meadows et al., 1972; 1st World Summit for the Environment, Stockholm; European Environmental Summit, 1972). At the European level, the first initiatives in the field of water took place at the same time. They were taken within the health-risk and competition-distortion umbrella, the environment being considered as a secondary issue (Bouleau & Richard, 2011). From 1986 onwards, the environment became a community competency with the adoption of the Single European Act.

According to various authors, three 'waves' can be identified in the history of European water policy (Aubin & Varone, 2002; Barraqué, 2004; Barraqué, Isnard, & Souriau, 2015; Bloech, 2004; Kaczmarek, 2006; Kallis & Nijkamp, 2000):

- A first wave of water directives (1973–86) aimed at reducing the health risks for citizens according to different water uses: bathing, drinking, and fish and shellfish production. Their focus was the establishment of water-quality norms and thresholds for different types of waters depending on their end use.
- A second wave of water directives (1987–92) aimed at reducing the most significant pressures that explained the poor chemical quality of aquatic ecosystems. As demonstrated by Kallis and Butler (2001), this wave opened up European water policy. With regards to governance, it positioned the public intervention of states between the wider public and a higher-level 'European referee', strengthening the consciousness of the public and its interest vis-à-vis water issues.
- A third wave (1993–onwards) recognized the need to strengthen the coherence and effectiveness of European water policy in order to achieve environmental objectives that had a clear ecological dimension. In addition, it gave the basis for addressing the issue of governance, the hydrological unit becoming the central scale for water governance and providing room for addressing local water-management issues. The cornerstone of this third wave was the European Union Water Framework Directive (WFD) adopted in the year 2000, the objective of which is good water status (including ecology) for all aquatic ecosystems. With the adoption of the WFD, water became (at that time) the environmental domain with the most developed European legislative framework (Kallis & Nijkamp, 2000). The adoption of the WFD was followed by the adoption of its two daughter directives on groundwater and hazardous substances. Finally, this was complemented by the adoption of the Floods Directive (2007) and the Marine Strategy Framework Directive (2008), two directives that followed the same (systemic) approach promoted by the WFD.

In conclusion, the logical frameworks that guide water management in Europe have shifted from fragmented and partial interventions to a coherent and systemic approach expected to deliver structural solutions to water-management problems. In the process, the European scale has had its part to play as a creator of basic principles while encouraging member states to apply the subsidiary principle to promote water management and action at more local scales.

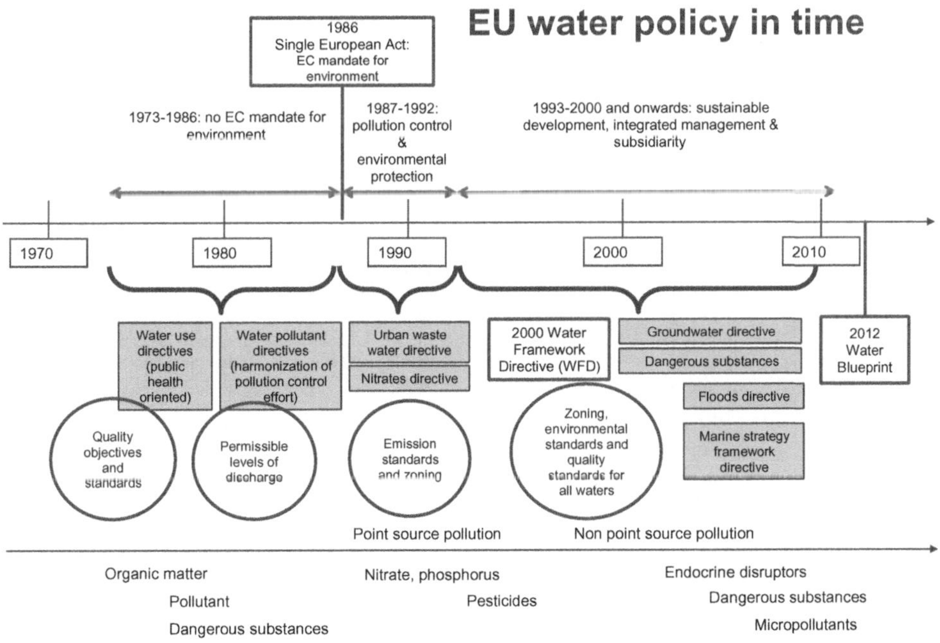

From 2014 on: towards enhancing local authorities and broadening water governance

Recent territorial reforms have empowered local authorities (municipalities and groupings of municipalities) with jurisdiction over water resource management and flood prevention (Box 5). The 2014 and 2015 Acts (MAPTAM and NOTRe) set out to reduce the very high number of local authorities involved in water and sanitation management in France (due to having no fewer than 36,000 municipalities, whilst the entire European Union has 89,000 combined). Groupings of municipalities of fewer than 15,000 inhabitants disappeared so that only larger contracting authorities remain in charge of flood prevention and water management. Following these laws, France will have fewer than 2000 water and sanitation service water authorities by 2020, reducing their number by around some 90%. From a state perspective these change consolidate the structuring of contracting authorities at administrative and catchment scale (SAGE).

Box 5. French water governance restructuring in light of decentralization, 1960–2016.

In France, since the 1960s, the governance of public action has changed. Governance used to be highly centralized with a regulator state that acted as the exclusive producer of policies. But nowadays governance has become more decentralized and involves more stakeholders in the co-construction of public action (Richard & Rieu, 2009). This is particularly true in the field of water, where both local authorities and the European institutions play crucial roles. European Union institutions influence norms and local authorities take the lead in designing and implementing local water policy. Today's governance has therefore become polycentric. The state maintains a key role in regulation, coordination and control, but is one of several stakeholders making policies, in close collaboration with local authorities, the private sector and non-governmental organizations (NGOs) (Richard et al., 2010).

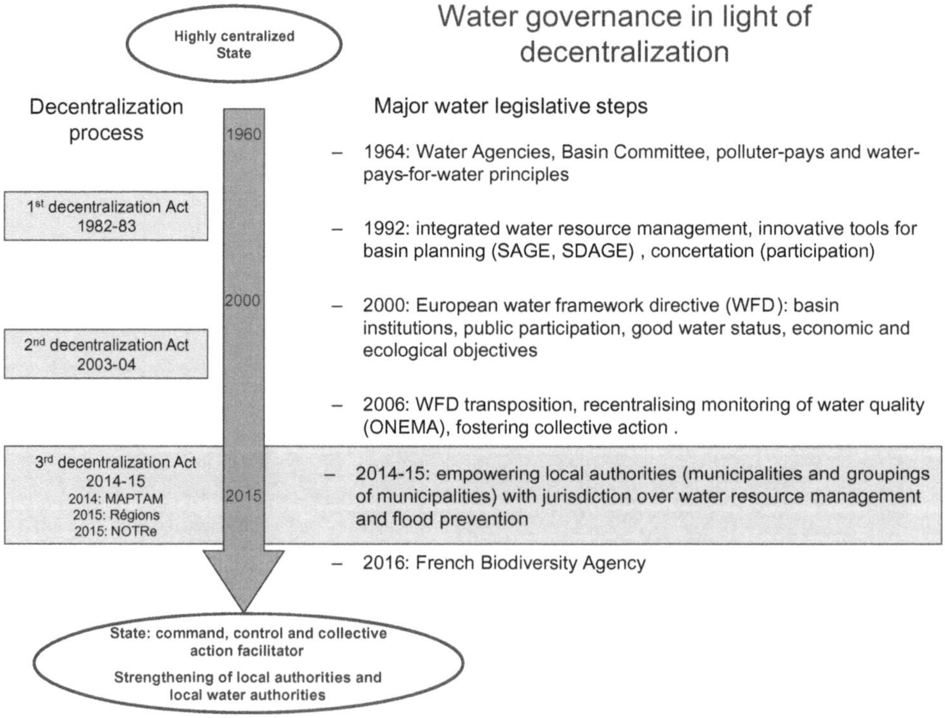

However, existing local water authorities, which are current contracting authorities at catchment scales, may be challenged. There is a recognized need to involve them fully in the current reforms. There is also a need to create local authorities in charge of developing ecological practices for river basin management for instance. *Départements*, which play an important role for rural areas, are seeking a new role to play – and one which the regulation has failed to specify. Regions received reinforced environmental responsibilities and are set to be leaders in biodiversity protection policy. The 2016 Biodiversity Act created new tools to bridge better water issues and other environmental concerns. ONEMA merged with three other public institutions to create the French Biodiversity Agency (FBA). As ONEMA is funded by water agencies, this provided a strong signal for enlarging the 'water agency system' to more global environmental policies. The new FBA is expected to work hand in hand with regions to foster innovative ways of addressing environmental issues, water included. Which prompts one to ask, of course: are the water agencies losing their central role?

Three key lessons can be drawn from the evolution of water resource governance during the past 50 years in France. First, water agencies have evolved from public-financing bodies setting out to protect and enhance water quality to planning and financing institutions, both of which are key in terms of supporting and implementing water policies. Today the creation of the FBA has enlarged the scope of the action of water agencies concerning terrestrial biodiversity.

Second, from the 1960s, the governance system shifted from a very centralized situation to a much more decentralized one (three decentralization acts from the 1980s until present). Since then, the state has been repositioning towards an increasingly polycentric governance system. Today, it is restricted to its position as national regulator in charge of implementing European directives (regulatory framework) vis-à-vis local authorities that have taken over a wider role in terms of implementing water management policies. Defining who the contracting authorities are remains a key issue in recent debates relating to ongoing territorial reforms.

A third significant change, and related to the previous one, is the opening of the water decision-making processes to the public. As a consequence, more stakeholders and interests are involved in/contribute to public policy-making. De facto, the more stakeholders, the more stakes and issues emerge: this in turn leads to an increasing need for integration and dedicated planning and financing tools, and in terms of governance, it can lead to the construction of co-responsibilities (Salles, 2009) and greater stakeholder commitment in the policy-making process.

Water resource governance has thus evolved towards more effectiveness, more efficiency and higher levels of both trust and engagement. Referring to the OECD Water Governance Principles, significant achievements have been made since 1964 with respect to: ensuring sound water management regulatory framework (principles 7 and 11) and promoting innovative water governance practices (principle 8); managing water at appropriate scales (basin, catchment, water territory), fostering coordination between scales and promoting stakeholder engagement (basin committee, local water commission) (principles 2 and 10); enhancing clear roles and responsibilities for water policy-making and regulation (state), operational management (local authorities, contracting authorities, local water authorities), and public financing authorities (water agencies, local authorities) (principles 1 and 6); enhancing data production and information, transparency (ONEMA, water agencies) (principle 5); and encouraging regular monitoring and evaluation driven by European norms (principle 12).

France's water-governance system is nowadays equipped with consistent institutional frameworks and tools. Remaining challenges relate to: encouraging policy coherence and efficiency through more effective cross-sectoral coordination (principle 3) and adapting the level of capacity of responsible authorities regarding their new roles following the last territorial reform (principle 4).

A synthesis of the evolution of French water governance

As exposed above, the OECD Water Governance Principles are used in this paper as a frame to analyse the evolution of French water governance. Table 2 exposes a synthesis of how France has implemented each principle since the 1960s. All principles have been adopted and enforced so far. However, various degrees of maturity in the enforcement are observed, as shown in Figure 3. Hence, the French system has not stopped evolving and still needs to evolve in the implementation of certain principles.

As shown in Figure 3 and Table 2, France has been evolving in the implementation of four principles:

- Principle 1: regarding the roles and responsibilities of local water authorities.
- Principle 3: by pushing water policy to be integrated into biodiversity policy.
- Principle 4: on strengthening the capacities of responsible authorities.
- Principle 8: in encouraging experimentations, e.g., regions as a coordinator of regional water, biodiversity, land planning and economic development policies.

Besides, France still needs to make progress in the implementation of certain principles, as showed in recent reforms:

- Principle 3: remains a challenge to reach greater inter-sectoral coherence (water and sector-based policies – agriculture, industry, urban and territorial planning etc.).
- Principle 10: on engaging citizens in the design and evaluation of water policies.
- Principle 11: regarding trade-offs across generations.

Changes dynamic: disputes as one of the main drivers

French water governance has changed within the consensual framework outlined above. Disputes were one of the main drivers of change. Disputes arise from different types of stakeholders. These stakeholders may be politicians, pressure groups, users and any civil society organizations. With an increasing role for the European Union's institutions, new mechanisms for dispute procedures opened. European citizens are entitled to refer a case to the European Court of Justice in the event of a member state failing to respect its obligations. Indeed, the French state, in keeping with others, has regularly been fined by the European Commission for failing to comply with various directives. For example, France was fined in 2014 for failing to implement European regulations in the fight

Table 2. Synthesis of how France implements the Organisation for Economic Co-operation and Development's (OECD) Water Governance Principles.

Aim	Water Governance Principles	Trends in the French water-governance system: Water service governance	Trends in the French water-governance system: Water resource governance
Effectiveness	Principle 1. Clearly allocate and distinguish the roles and responsibilities for water policy-making, policy implementation, operational management and regulation, and foster coordination across these responsible authorities	Clear responsibilities defined by laws and decrees to each stakeholder involved in water service governance since the 1960s.	From a centralized political system in the 1960s to a more decentralized system today. Towards a clarification of roles and responsibilities with the current territorial reforms: • Local authorities: contracting authorities, financing, towards more competencies • Water agencies: from purely financial agencies in the 1960s to planning and financial water agencies in the 1990s and toward planning and financial agencies for biodiversity today • State: regulation and control Today: clear responsibilities: regulation (state), operational management (local authorities, contracting authorities) and public financing authorities (water agencies, local authorities)
	Principle 2. Manage water at the appropriate scale(s) within integrated basin governance systems to reflect local conditions and foster coordination between the different scales	More opportunities with recent reform to have local authorities managing the full water cycle. Fostered coordination between scales: basin, catchment, water territory, administrative territories	
	Principle 3. Encourage policy coherence through effective cross-sectoral coordination, especially between policies for water and the environment, health, energy, agriculture, industry, spatial planning and land use	Encouraging policy coherence and efficiency through more effective cross-sectoral coordination (biodiversity, agriculture, urban planning etc.): evolution underway spurred by recent reforms	
	Principle 4. Adapt the level of capacity of responsible authorities to the complexity of the water challenges to be met, and to the set of competencies required to carry out their duties	Recent reforms aim at increasing the capacity of responsible local authorities for water and sanitation services by increasing their size	Strengthened capacity of the responsible authorities Further capacity-building is required following last territorial reform

(*Continued*)

Table 2. (Continued).

Aim	Water Governance Principles	Trends in the French water-governance system: Water service governance	Trends in the French water-governance system: Water resource governance
Efficiency	Principle 5. Produce, update and share timely, consistent, comparable, and policy-relevant water and water-related data and information, and use it to guide, assess and improve water policy	Recent reforms aim to enhance data production and information and transparency with the creation of the French Agency for Biodiversity. It manages the national information system on water and sanitation services, and data on water resources monitoring	
	Principle 6. Ensure that governance arrangements help mobilize water finance and allocate financial resources in an efficient, transparent and timely manner	Water agencies' fees on water bills fund investments and resources-protection measures 'Water pays for water' principle implies that expenditures by water and sanitation services are funded by water bills and dedicated grants or loans. The intra-basin solidarity and the water pays for water principle, in place since 1964, are today threatened by a state deduction from the water agencies' budget A strict principle of cost recovery is imposed to services Remaining questions about the ability to fund the replacement of infrastructures	
	Principle 7. Ensure that sound water management regulatory frameworks are effectively implemented and enforced in pursuit of the public interest	Ancient, robust and efficient but complex regulatory framework	
	Principle 8. Promote the adoption and implementation of innovative water-governance practices across responsible authorities, levels of government and relevant stakeholders	Some recent advances on the water-governance framework promoting social learning, encouraging experimentation (e.g., regions) and synergies across sectors and scales	
Trust and engagement	Principle 9. Mainstream integrity and transparency practices across water policies, water institutions and water-governance frameworks for greater accountability and trust in decision-making	Improved trust and engagement through improved performance of water and sanitation services. Still progress to be made involving water users and communicating better with citizens	Legal and institutional frameworks that hold decision-makers and stakeholders accountable (right to access information, e.g., water agencies; adoption of multi-stakeholder approaches etc.)
	Principle 10. Promote stakeholder engagement for informed and outcome-oriented contributions to water policy design and implementation	Promoted stakeholder engagement: basin committee, local water commission, user committees etc.	
	Principle 11. Encourage water-governance frameworks that help manage trade-offs across water users, rural and urban areas, and generations	Urban–rural cooperation and solidarity enhanced by recent reforms encouraging the groupings of local authorities	Water-governance framework promoting participation, empowering local authorities and users Further attention to be paid to trade-offs across generations
	Principle 12. Promote regular monitoring and evaluation of water policy and governance where appropriate; share the results with the public and make adjustments when needed	Regular policy evaluation by the 'Cour des Comptes', parliament or senate available to the public	Encouraged regular monitoring and evaluation driven by European norms

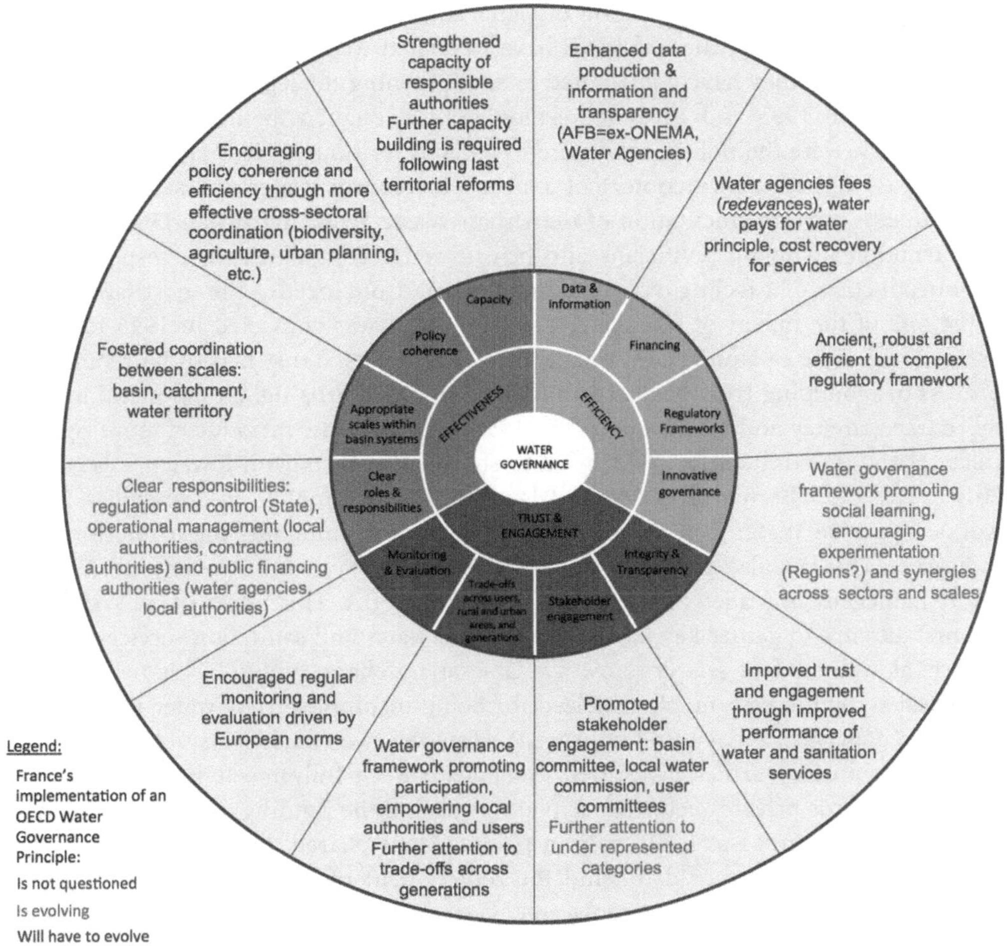

Figure 3. Recent and needed evolutions in the implementation of the Organisation for Economic Co-operation and Development's (OECD) Water Governance Principles in France.

against nitrate pollution. These cases provide added opportunities for environmental non-governmental organizations (NGOs) to be heard by the government. There is no specific policy evaluation body, but various public institutions are entitled to assess environmental policies. Disputes may be triggered by events representative of failures of the system.[7]

Two emblematic disputes in the sector that led to governance change are set out here for illustrative purposes – the first in the realm of water services, the second in the field of water resource management.

Disputes in water and sanitation utility management: the way to local authority capacity reinforcement

As already explained, water and sanitation services are characterized in France by the heavy involvement of private companies. This private sector participation has been debated in France since the creation of water and sanitation services at the end of the 19th century. However, these debates have never led to any change in the delegation model. That said, they have contributed to strengthening the legal framework. The laws passed between 1993 and 1995, for example, and designed to increase integrity and transparency, were the political response to a major legitimacy crisis faced by the sector for the past 50 years. In a context of a strong increase in water and sanitation tariffs partly due to the implementation of the Urban Waste Water European Directive, cases of corruption involving politicians and private water companies were responsible for the introduction of a feeling of suspicion on the part of citizens. The most famous case is the one of the mayor of Grenoble, Carignon, who was convicted in 1995 for illegal use of the private operators', Lyonnaise des Eaux, funds. Trust had to be rebuilt. The process of rebuilding trust has been unfolding since towards tighter competition rules, more transparency and information. The 1993, the Sapin law introduced more rigorous rules relating to competition, which led to shorter contracts with lower prices (Colon, 2017). The 1995 Barnier and Mazaud laws established annual accountability frameworks. The 2006 Water Act introduced compulsory accountability on key performance indicators. This law also created a national information system on water and sanitation utility management (called SISPEA), and run by the FBA. This information system was the first attempt to gather key factual material on water and sanitation services. Annual reports as well as data sets are now available at no charge online. This new national statistical system is very much criticized for being incomplete (few water utilities send data, and data quality has to be improved). However, we consider its very existence as progress, requiring further investments to make sure it fully meets its role. A central criticism towards private-sector participation was that the funding of investments made by private companies was far from transparent. Many shared the view that even when private companies were paid to fund the replacement of existing equipment, private operators tended to postpone replacements right until the end of their contracts. Contracts were drafted so that responsible authorities were not entitled to be paid back for what had not been invested. The 2006 Water Act also contributed to clarifying this by making such financial arrangements illegal (Guérin-Schneider & Colon, 2017). Since 2006, if a planned replacement has not been carried out by the end of the contract, private operators are obligated to reimburse the authority. Even the recent decentralization phase III law (NOTRe 2015) may be read as a consequence of this legitimacy crisis of the 1990s, essentially linked with the weakness of local authorities to ensure their accountability in relation to consumers and citizens alike.

After the remunicipalization of the Parisian water utility, some stakeholders considered that the debate on private-sector participation might well return.[8] In fact, the opposite happened. By showing clearly the capacity and responsibility of local authorities to choose freely the way they want the service to be operated, and that such choices were reversible, this contributed to a shift in the debate. The issue was not whether the private sector should operate water utilities. The real issue relates to the capacity of responsible authorities to manage their relationships with their operators properly in all cases (public, public/private or private operators). This includes having the capacity to collect data and analyse them, control private operations and design orientations for a sustainable future (Roche, Guerber, et al., 2016). Good practices of performance contracts with no regard for the public or private status of the operator are now shared, for example, in ASTEE congresses, gathering representatives from all kind of systems (Roche, Le Fur, & Canneva, 2012).

Disputes in water resource management: the example of the reform of users' representation in basin committees

Basin committees, which were created in the mid-1960s, gather public and private stakeholders concerned by water management in a river basin. They debate and define through dialogue processes the main lines of local water resource management. Their role is that of a 'water parliament'. The Environment Code states there are three categories of members: category 1: representatives of local authorities (regions, *départements* and municipalities) which account for 40% of total members; category 2: water users (domestic users, professional organizations, NGOs for environmental protection or users defence, fisheries, experts) (40% of total members); and category 3: representatives of the state (20% of total members). The committee is headed by a president elected by a vote of and among category 1 and 2 members.

With growing environmental awareness from the 1980s, the composition of these river committees has started to be regularly contested. In 2013, the French government organized the Second Conference on the Environment that brought together stakeholders from the state, local representatives and civil society to discuss and challenge French environmental policies. The need to work on water governance became evident and strongly felt. A taskforce was created with the National Water Committee on Governance in 2013. The main dispute, expressed by NGOs, was that river committees suffer from 'too big a representation of farmers, with little room for domestic users'.[9]

A clear obstacle to change was that the mandates of the various basin committee members were about to terminate (June 2014) when the French national water committee ('Comité national de l'eau') made its proposals (at the end of 2013). Given that it takes between 4 and 6 months to renew memberships, there was insufficient time to dwell upon deep regulatory reform. The decision was finally made to issue a decree by the Ministry of Ecology, Sustainable Development and Energy on 27 June 2014. This text divides the category 2 of members between economic and non-economic users (domestic, NGOs). Two subcategories of professional representatives were created: 'Agriculture, fishery, fish farming, canal transport and tourism' professionals and 'Industries and cottage industries'.

This decree was the first step towards deeper changes relating to local water governance. As the 'Cour des Comptes" 2015 report states, despite this 2014 reform, civil society representation remains low. Farmer representation, by way of contrast, is mainstreamed by very influential organizations, and local authority representation is questioned due to decentralization laws. So while the existence of river committees has not been called into question, there is manifestly room for change in line with society's evolving expectations.

It is clear, then, that France's water-governance system has evolved quite consistently during the course of the last 50 years. In the realm of water resource management, the governance framework has become more inclusive, with an increasingly important role to civil society, NGOs and with constantly reinforced participatory processes. The influence of Europe has led France to move towards more result-oriented policies.

France is today moving to integrate water in sectorial policies. In the water services realm, the governance system has moved towards more power to responsible authorities with more integrity and transparency, and an enhanced role to civil society.

Discussion of the model of change compared with the OECD water governance cycle

France's water-governance system has been presented in this paper as shared between water resource management, on the one side, and water and sanitation services, on the other. Such a presentation makes sense in that these two water worlds have developed over time as two connected but parallel sets of actorhood, rules, funding and territories. However, the focal point that makes these two worlds now meet is the issue of who are the contracting authorities. Who is entitled and owns the capacity to fund, organize and act for water management? On the side of water resource management, there is a need to develop a responsible, able and empowered authority to manage flood risk. On the side of water and sanitation utilities, there is clearly a need to strengthen responsible authorities to build a sustainable future within the context of increasing uncertainties and lack of public finance. This question of the capacity of the responsible authority is the core issue regarding the current French water-governance system being addressed by recent laws.

French water governance is consistent with OECD principles. The current legislation is based on a 50-year experience of IWRM and more than a 100-year experience of public/private partnerships for water and sanitation utility management. Worthy of note is that the framework defined 50 years ago has remained largely consensual. The state has played a key role in the building of a common culture of water governance, implemented and executed by its state engineers.

How has such an evolution occurred? We are far from having a 'rational' process of policy-making defined in the water policy cycle by the OECD: policy formulation/implementation/monitoring/evaluation/new policy formulation etc. Constant disputes have been responsible for leading to movement within this framework. These disputes arise from stakeholders who strive to ensure that their perspective is taken into account in the formulation of policy. Evaluations occur occasionally on a topic when the need to tackle a specific issue is felt, perhaps as expressed by a pressure group. These evaluation reports may be held by politicians ('*rapports parlementaires*') or by an independent public finance control agency ('Cour des Comptes') or by advisory board proposals placed near the ministries (CGEDD). Thus, there is not one state body in charge of water policy assessment, but a diversity of public institutions that can produce evaluation reports within their mandate. These observations are consistent with advances in policy analysis theories (e.g., Sabatier & Schlager, 2000).

Lessons from France

This paper had two purposes: to present recent developments within the French water-governance system, and to learn from its evolution to understand the changing dynamics.

The main idea advocated here is that the water-governance framework that is well known worldwide has attained a state of consensus in France. However, the model is constantly challenged, which itself encourages change. Recent developments have already been criticized. The compulsory transfer of the responsibility of water and sanitation services to groupings of municipalities may generate a less flexible and grounded management (Barbier, 2015). In February 2017, new legislation was introduced to remove this obligation. The limitation of this changing dynamic model is that there is little space for less powerful actors. In other words, it is the powerful actors who retain the upper hand.

What is there to learn, then, from France's system of water governance? In fact, there are two key points:

- The first is the question of who bears the role of contracting authorities – a crucial issue in the French context. The French feedback on the 'who does what?' issue illustrates that it is clearly inherited from the country's complex institutional setting, which goes far beyond the water-governance system alone.
- Second, it is imperative to ask what is the place of a water-governance system in general. After 50 years of specific water policy, it is beginning to fragment at the edges and face new challenges. There is a need to integrate issues relative to water into sectoral policies such as urban planning etc. Furthermore, water is no longer the 'crown jewel' of an emerging environmental policy. Other environmental issues such as atmospheric pollution, climate change and biodiversity are now competing for the centre stage. The creation of the FBA in 2017 seeks to integrate water governance with broader biodiversity management issues. France's water-governance system nevertheless benefits from a strong and well-established basis that has proven to be very resilient over time. It has demonstrated sufficient strength and flexibility to be able to adapt and to play a key role in broader environmental policy issues, too.

Through this French case, we highlighted how the main OECD Water Governance Principles may be implemented. They appear necessary, but their sole implementation may not be enough to assess the quality of a national water policy. This would require a specific policy-evaluation analysis to assess the results and outcomes of the French water policy. Our paper stresses the main limits and advances in the French water-governance system of which the authors are aware, but it does not provide a thorough policy evaluation.

Is this model replicable? We do not believe in one-size-fits-all models. We showed how much the French water governance is bound tightly to the organization of the state, local authorities and private companies. Some principles may be and have been replicated, as the delegation model or the river basin management of water resources. Beyond that, each country has to find solutions to adjust to their own local context.

Notes

1. ASTEE, the Association Scientifique et Technique pour l'Eau et l'Environnement, is a French non-governmental organization (NGO) bringing together water professionals and researchers to work on, share and produce knowledge on environmental issues.

2. OECD, see http://www.oecd.org/governance/oecd-principles-on-water-governance.htm/.
3. For simplification, this paper focuses on Metropolitan France, excluding overseas territories (for a specific review, see Roche, Colas-Berlcour, et al., 2016).
4. French legislation is meant to comply with European Union regulation. Since 2016, for instance, the European Directive on concession has been enforced and applied to the water and sanitation sector.
5. Note the permanent dialectical position between those seeing water agencies as 'instruments to enforce the polluter-pays principle' (chiefly, neoclassical environmental economists) and those putting an emphasis on cost sharing and common resources management (political scientists). In 1964, economists put forward the eco-tax model in the debates on the Water Act (Bouleau & Richard, 2011). The principle of fees paid by water users was accepted only provided that it be used for investment. The elected representatives insisted that the rate should not be too high at the outset in order to avoid having a surplus which might be used for other purposes. The compromises resulted in setting up water fees that, in fact, hardly are 'environmental taxes' and rather are an intermediate between economic efficiency and collective savings.
6. ONEMA, the Office national de l'eau et des milieux aquatiques, is a state public body under the authority of the Ministry of Environment, now the French Biodiversity Agency (Agence Française pour la Biodiversité).
7. For instance, storm Xynthia of 2010 that hit the west coast of France caused 59 deaths and €1.5 billion of damage. Local authorities and the state were criticized for having allowed urbanization on submersible coastal lands, exposing thousands to a flood risk in so doing.
8. The expected remunicipalization wave after the Paris case has been very limited. From 2010 to 2014, the rate of the population served by water public operators has stagnated around 40% according to the ONEMA.
9. Coordination eau Adour Garonne, letter to the committee president (5 December 2014)

Disclosure statement

No potential conflict of interest was reported by the authors.

ORCID

Marine Colon http://orcid.org/0000-0001-7302-591X
Sophie Richard http://orcid.org/0000-0001-8969-1634

References

Agences de l'Eau. (2017). *Les leviers d'action des agences de l'eau pour la gestion durable de l'eau.* Agences de l'Eau. Retrieved from http://www.lesagencesdeleau.fr/les-agences-de-leau/les-leviers-daction-des-agences-de-leau/

Aubin, D., & Varone, F. (2002). *European Water Policy: A path towards an integrated resource management?* (pp. 28). Louvain-La-Neuve: EUWARENESS. AURAP-UCL.

Barbier, R. (2015). Le modèle institutionnel de l'eau potable au défi de sa durabilité: Enjeux, acteurs et dynamiques de rationalisation en France métropolitaine. *Politiques Et Management Public, 32*(2), 129–145. doi:10.3166/pmp.32.129-145

Barone, S., Dedieu, C., & Guérin-Schneider, L. (2016). Le suppression de l'ingénieriee publique de l'Etat dans le domaine de l'eau: Les effets paradoxaux d'une réforme néo-managériale. *Politiques Et Management Public, 33*(1), 49–67. doi:10.3166/pmp.33.49-67

Barraqué, B. (2004). Aspects institutionnels, socio-économiques et juridiques de la gestion durable de l'eau en Europe. In Université d'Artois (Ed.), *Actes de la journée d'études* (pp. 167–172). Arras: Université d'Artois.

Barraqué, B., Isnard, L., Barbier, R., & Canneva, G. (2011). *Trajectoires techniques et institutionnelles des services d'eau en Europe de l'ouest, aux Etats-Uis et en Australie* Recherche No. 5.1 22. Paris: ANR Eau & 3E.

Barraqué, B., Isnard, L., & Souriau, J. (2015). How water services manage territories and technologies: History and current trends in developed countries. In Q. Grafton, K. A. Daniell, C. Nauges, J.-D. Rinaudo, & N. W. W. Chan (Eds.), *Understanding and managing urban water in transition* (Vol. 15, pp. 33–59). Dordrecht: Springer Netherlands. doi:10.1007/978-94-017-9801-3_2

Barraqué, B., & Laigneau, P. (2017). Agences de l'eau: Rétrospection prospective. *Responsabilité & Environnement, 87*, 114–120.

Bloech, H. (2004). European water policy and the Water Framework Directive: An overview. *Journal for European Environmental & Planning Law, 1* (3)

Bouleau, G. (2007). La gestion française des rivières et ses indicateurs à l'épreuve de la directive cadre. Analyse néo-institutionnelle de l'évaluation des cours d'eau en France (Thèse de doctorat en Sciences de l'environnement). Paris: AgroParisTech – ENGREF.

Bouleau, G., & Richard, S. (2011). *French water legislation within the context of the framework directive. Recent developments.* Sao Paulo: Anna Blume.

Canneva, G. (2012). Les modèles de régulation des services d'eau et d'assainissement. In P.-A. Roche & G. Canneva (Eds.), *Améliorer la performance des services publics d'eau et d'assainissement* (pp. 51–54). Paris: ONEMA .

Clark, E., & Mondello, G. (1999). Institutional constraints in water management: The French case. *Water International, 24*(3), 266–268. doi:10.1080/02508069908692170

Colon, M. (2017). *Observatoire des services publics d'eau et d'assainissement Impacts des procédures de mise en concurrence dites « Loi Sapin » sur les services d'eau et d'assainissement en 2014.* Paris: Agence Française pour la Biodiversité.

Crespi-Reghizzi, O. (2014). Providing a municipal infrastructure: How did Paris and Milan finance their water and sanitation infrastructure (1853–1925)? *Flux – Cahiers Scientifiques Internationaux Réseaux et Territoires.*

Drobenko, B., & Fromageau, J. (Eds.). (2015). *La loi sur l'eau de 1964: Bilans et perspectives.* Paris: Editions Johanet.

Guérin-Schneider, L., & Colon, M. (2017). Le développement de pratiques d'accountability dans le secteur des services d'eau: Un changement institutionnel inabouti? Presented at the Congrès annuel de l'Association Francophone de Comptabilité, Poitiers.

Gurría, A. (2009). Sustainably managing water: Challenges and responses. *Water International, 34*(4), 396–401. doi:10.1080/02508060903377601

Kaczmarek, B. (2006). *Un nouveau rôle pour les agences de l'eau? Essai pour une politique franco-européenne de l'eau rénovée.* Paris: Johanet.

Kallis, G., & Butler, D. (2001). The EU water framework directive: Measures and implications. *Water Policy, 3*, 125–142. doi:10.1016/S1366-7017(01)00007-1

Kallis, G., & Nijkamp, P. (2000). Evolution of EU water policy: A critical assessment and a hopeful perspective. *Zeitschrift für Umweltpolitik und Umweltrecht, 3*, 301–335.

Levraut, A.-M., Payen, D., Coppinger, N., Cholley, F., Madignier, M.-L., Bénézit, J.-J., … Laganier, R. (2013). *Evaluation de la politique de l'eau (Rapport d'analyse)* Retrieved from. Paris: Ministère de l'écologie, du développement durable et de la mer. http://www.ladocumentationfrancaise.fr/var/storage/rapports-publics/134000639.pdf

Meadows, D. H., Meadows, D. L., Randers, J., & Behrens III, W. W. (1972). *The limits to growth.* New York: Universe Books.

Molle, F. (2012). La GIRE: Anatomie d'un concept. In Presses de l'Université de Québec (Ed.), *Gestion intégrée des ressources en eau: Paradigme occidental, pratiques africaines* (pp. 23–53). Québec: Presses de l'Université de Québec.

Nicolazo, J.-L. (1993). *Les Agences de l'Eau.* Paris: Pierre Johanet et Fils Editeurs.

ONEMA. (2016). *Observatoire des services publics d'eau et d'assainissement. Panorama des services et leur performance en 2013* (pp. 149). Paris: ONEMA.

Peyrou, D., & Roche, P.-A. (2006). Implication des divers acteurs à des échelles de territoires emboîtées. In *Le développement durable, c'est enfin du bonheur*. Cerisy: Les éditions de l'Aube.

Pezon, C. (1999). La gestion du service de l'eau en France. Analyse historique et par la théorie des contrats (1850 à 1995). (Thèse de doctorat en sciences de gestion, sous la direction de Raymond Leban). Paris: CNAM.

Richard, S., Bouleau, G., & Barone, S. (2010). Water governance in France. Institutional framework, stakeholders, arrangements and process. In P. Jacobi & P. Sinisgali (Eds.), *Water governance and public policies in Latin America and Europe* (pp. 137–178). Sao Paulo: Anna Blume.

Richard, S., & Rieu, T. (2009). Vers une gouvernance locale de l'eau en France: Analyse d'une recomposition de l'action publique à partir de l'expérience du schéma d'aménagement et de gestion de l'eau (SAGE) de la rivière Drôme en France. *VertigO. 9*, 1. Retrieved from http://vertigo.revues.org/index8306.html

Roche, P.-A. (2002). Les institutions françaises face à la directive-cadre européenne sur l'eau. *Responsabilité & Environnement, 25*, 75–90.

Roche, P.-A. (2005). *Besoins de recherche liés à l'application de la directive-cadre européenne sur l'eau*. Paris: Elsevier.

Roche, P.-A., Colas-Berlcour, F., Vial, J.-C., & Tandonnet, M. (2016). *Propositions pour un plan d'action pour l'eau dans les régions et départements d'outremer et à Saint-Martin* (La documentation française) 234. Paris: CGEDD-IGA-CGAAER. Ministère de l'écologie, du développement durable et de l'énergie; Ministère des Outre-Mer.

Roche, P.-A., Guerber, F., Nicol, J.-P., & Simoni, M.-L. (2016). *Eau potable et assainissement: à quel prix ?* (No. 10151–1) 561. Paris: CGEDD-IGA. Ministère de l'Environnement, de l'Energie et de la Mer, Ministère de l'Intérieur.

Roche, P.-A., Le Fur, S., & Canneva, G. (2012). *Improving the performance of water and sanitation public services*. Nanterre: ASTEE.

Sabatier, P. A., & Schlager, E. (2000). Les approches cognitives des politiques publiques: Perspectives américaines. *Revue Française De Science Politique, 50*(2), 209–234. doi:10.3406/rfsp.2000.395465

Salles, D. (2009). Environnement: La gouvernance par la responsabilité? *VertigO* (Hors Série 6). Retrieved from http://vertigo.revues.org/9179 10.4000/vertigo.9179

SOeS. (2015). *Chiffres clés de l'environnement*. Repères, Édition 2015. Paris: Service de l'observation et des statistiques. Commissariat général au développement durable.

Valdovinos, J. (2012). The remunicipalization of Parisian water services: New challenges for local authorities and policy implications. *Water International, 37*(2), 107–120. doi:10.1080/02508060.2012.662733

Index

Note: Page numbers in *italic* type refer to figures
Page numbers in **bold** type refer to tables
Page numbers followed by 'n' refer to notes

For Product Safety Concerns and Information please contact our EU representative GPSR@taylorandfrancis.com Taylor & Francis Verlag GmbH, Kaufingerstraße 24, 80331 München, Germany

Batch number: 10726412

Printed by Printforce, the Netherlands